Linas Urbonas

One-atom-maser

Linas Urbonas

One-atom-maser

Experiments

Südwestdeutscher Verlag für Hochschulschriften

Impressum/Imprint (nur für Deutschland/ only for Germany)
Bibliografische Information der Deutschen Nationalbibliothek: Die Deutsche Nationalbibliothek verzeichnet diese Publikation in der Deutschen Nationalbibliografie; detaillierte bibliografische Daten sind im Internet über http://dnb.d-nb.de abrufbar.

Alle in diesem Buch genannten Marken und Produktnamen unterliegen warenzeichen-, marken- oder patentrechtlichem Schutz bzw. sind Warenzeichen oder eingetragene Warenzeichen der jeweiligen Inhaber. Die Wiedergabe von Marken, Produktnamen, Gebrauchsnamen, Handelsnamen, Warenbezeichnungen u.s.w. in diesem Werk berechtigt auch ohne besondere Kennzeichnung nicht zu der Annahme, dass solche Namen im Sinne der Warenzeichen- und Markenschutzgesetzgebung als frei zu betrachten wären und daher von jedermann benutzt werden dürften.

Verlag: Südwestdeutscher Verlag für Hochschulschriften Aktiengesellschaft & Co. KG
Dudweiler Landstr. 99, 66123 Saarbrücken, Deutschland
Telefon +49 681 37 20 271-1, Telefax +49 681 37 20 271-0
Email: info@svh-verlag.de
Zugl.: München, LMU, Diss., 2009

Herstellung in Deutschland:
Schaltungsdienst Lange o.H.G., Berlin
Books on Demand GmbH, Norderstedt
Reha GmbH, Saarbrücken
Amazon Distribution GmbH, Leipzig
ISBN: 978-3-8381-1288-6

Imprint (only for USA, GB)
Bibliographic information published by the Deutsche Nationalbibliothek: The Deutsche Nationalbibliothek lists this publication in the Deutsche Nationalbibliografie; detailed bibliographic data are available in the Internet at http://dnb.d-nb.de.

Any brand names and product names mentioned in this book are subject to trademark, brand or patent protection and are trademarks or registered trademarks of their respective holders. The use of brand names, product names, common names, trade names, product descriptions etc. even without a particular marking in this works is in no way to be construed to mean that such names may be regarded as unrestricted in respect of trademark and brand protection legislation and could thus be used by anyone.

Publisher: Südwestdeutscher Verlag für Hochschulschriften Aktiengesellschaft & Co. KG
Dudweiler Landstr. 99, 66123 Saarbrücken, Germany
Phone +49 681 37 20 271-1, Fax +49 681 37 20 271-0
Email: info@svh-verlag.de

Printed in the U.S.A.
Printed in the U.K. by (see last page)
ISBN: 978-3-8381-1288-6

Copyright © 2010 by the author and Südwestdeutscher Verlag für Hochschulschriften Aktiengesellschaft & Co. KG and licensors
All rights reserved. Saarbrücken 2010

Contents

1 Introduction   **7**

2 Theoretical Background   **11**
    2.1 Basics 11
    2.2 Jaynes-Cummings Model 16
    2.3 Master Equation 17
    2.4 Steady State 19
    2.5 Vacuum Rabi Oscillations 19

3 Experimental Setup   **23**
    3.1 Overview 23
    3.2 Atomic Beam Production 25
    3.3 Excitation 29
    3.4 Detection 33
    3.5 Resonator 39
    3.6 ^{3}He cryostat 45

4 Measurements with the Atomic Beam   **51**
    4.1 Magnetic Field Compensation 51
    4.2 Resonance Curve of the Maser 53
    4.3 Velocity Selection and TOF Measurements 57
    4.4 Vacuum Rabi Oscillations 61

5 Laser System and Spectroscopy   **69**
    5.1 Overview 69
    5.2 Laser Frequency Stabilization 72
    5.3 First Stage 76
        5.3.1 The D2 Transition of Rb 76
        5.3.2 Doppler-free Saturation Spectroscopy 77
    5.4 Second Stage 83
    5.5 Third Stage 87

6	**Magnetic Field Compensation**	**91**
	6.1 Overview	91
	6.2 High Resolution Magnetic Field Compensation	95
	6.3 Velocity Selected MFC	98

Bibliography   **109**

List of Figures

2.1	Classical field in a metallic cavity	12
2.2	The main principle of the micromaser	14
2.3	The theoretical model of micromaser	16
2.4	Jaynes-Cummings model for micromaser	17
2.5	Relaxation processes in the micromaser	18
2.6	Steady state photon distribution in the micromaser	20
2.7	Theoretical Rabi oscillations	21
3.1	Main experimental setup	24
3.2	Atomic oven	26
3.3	The speed probability density functions	27
3.4	Atomic beam collimator with the excitation region	29
3.5	New three-step diode laser setup	31
3.6	Test of the stability of the system	32
3.7	Plateau curve of the channeltron	35
3.8	Channeltron box	36
3.9	Field-ionization signal for atoms	38
3.10	Auxiliary detection system	39
3.11	Resonator	40
3.12	The transverse electric field distribution	41
3.13	Surface resistance dependence on the temperature	42
3.14	Resonator frequency dependence on the applied piezo voltage	43
3.15	Resonator Q-factor measurement setup	44
3.16	Resonator Q-factor measurement	45
3.17	Mean thermal photon number dependence on the temperature	46
3.18	^{3}He cryostat	48
4.1	Magnetic field compensation scheme	52
4.2	Experimental magnetic field compensation curves	52
4.3	Resonance curve of the maser	54
4.4	The experimental resonance curve measurements for different count rates	55
4.5	The resonance curve linewidth dependence on the count rate	56

4.6	The maser line resonance shift dependence on the count rate	57
4.7	Time of flight measurement scheme	58
4.8	Time of flight measurement for two different Stark voltages .	60
4.9	Time of flight calibration measurement	61
4.10	Measurement scheme of vacuum Rabi oscillation	62
4.11	Experimental results of vacuum Rabi oscillation measurement 1	63
4.12	Experimental results of vacuum Rabi oscillation measurement 2	64
4.13	New measurement scheme of Rabi oscillation	65
4.14	Experimental results of vacuum Rabi oscillation measurement 3	66
4.15	Experimental results of vacuum Rabi oscillation measurement 4	67
5.1	Excitation scheme	70
5.2	Diode laser system	71
5.3	Laser frequency stabilization scheme	73
5.4	Step function response	75
5.5	Energy manifolds of Rb transitions	77
5.6	Setup of Doppler-free saturation spectroscopy	79
5.7	Hyperfine transitions for the first stage	81
5.8	Setup of absorption spectroscopy	84
5.9	Spectroscopy signal for the 2−nd stage laser locking	86
5.10	The signal from the atomic beam	87
5.11	The test of the laser stabilization	89
6.1	The main setup of magnetic field compensation in one-atom maser	92
6.2	Quantum mechanical explanation of magnetic field compensation in one-atom maser	93
6.3	The experimental results of magnetic field compensation . . .	94
6.4	High resolution magnetic field compensation scheme	96
6.5	Experimental results of high resolution magnetic field compensation	97
6.6	The scheme of experimental coherent atomic control realization	100
6.7	Theoretical fit and experimental results	101
6.8	Experimental results of magnetic field scan	102
6.9	Magnetic field scan for different classes of atomic velocities .	103
6.10	The scan of magnetic field with the higher first stage laser power	104
6.11	Scheme of the theoretical model for the Hanle precession calculation	105
6.12	Hanle precession over a large range of magnetic field scan . .	106

List of Tables

2.1   Main micromaser parameters   15

3.1   Comparison of the old and the new one laser systems   30
3.2   The main parameters of the corresponding diode laser stages   31

4.1   Main experimental parameters   53

6.1   General expression for the excitation probability coefficients
      for different excitation schemes 107

ness
LIST OF TABLES

Chapter 1
Introduction

Across a broad front in physics, cavity quantum electrodynamics (CQED) plays an important role as a fundamental system and research field where light-matter interaction is investigated. It covers a wide spectrum of aspects ranging from fundamental studies of pure number state generation to quantum information processing.

The birth of this field of research was introduced by the publication of Edward Mills Purcell in 1946, where he noticed that the rate of spontaneous emission of an atom can be significantly enhanced by coupling it to an electrical circuit resonant with the atomic radio-frequency transition [Pur46].

CQED study electromagnetic field in a confined space and the radiative properties of atoms in such field. And obviously, one of the simplest systems in the scope of CQED is a single atom interacting with a single field mode. The theory describing this system was first developed by Jaynes and Cummings in 1963 [JC63], providing the most basic model to CQED experiments.

The main obstacles in the experimental realization of a coupled atom-cavity system is that the coupling is disturbed by spontaneous emission of the atom and damping of the cavity field. Thus, the most relevant regime for observing quantum phenomena of the atom-field dynamics is reached in the schemes where the coupling is strong enough to exceed the spontaneous emission rate as well as cavity damping. Fortunately, the tremendous progress in controlling single atoms as well as in manufacturing high finesse cavities, accomplished over the last decades, has allowed experimentalists to achieve the strong coupling with single atoms in cavities in two distinct regimes, the microwave and the optical region.

At microwave frequencies, highly excited Rydberg atoms are coupled to the field of a superconducting cavity with a very high Q-factor while crossing the cavity mode one by one [WVEB06][HR06]. Strong coupling in the optical domain has been reached using ultra-cold ground-state atoms and cavities with small mode volume [MN$^+$05][RFH$^+$03].

In recent years, cavity experiments have also been conducted on a variety of solid-state systems resulting in many interesting applications. Some outstanding examples of these are microlasers [ACDF94][FYYH+06], photon bandgap structures [FSH+05] and quantum dot structures [MKB+00] in cavities.

In experimental realization of such CQED systems in microwave regime, two approaches have been proven to be particularly successful. One was pioneered by our group based on the original idea and continuing input of H. Walther, which utilized closed resonators. Meanwhile, the other one was developed in Paris by S. Haroche using open resonators.

In this thesis, an atom-cavity system in a microwave regime is investigated. A one-atom-maser or micromaser provides the possibility of making a detailed study of the atom-field interaction in closed cavities. Provided the atomic and cavity decay is negligible, theory predicts a coherent interaction between the atom and the field which produces a sinusoidal oscillations of the population known as Rabi oscillations. These oscillations are governed by atom-field coupling constant g. Strong coupling regime in this case is characterized by g being much larger than atomic decay constant γ and cavity decay constant κ: $g >> \gamma, \kappa$.

To satisfy this condition, these parameters should be chosen correspondingly. In modifying coupling constant g, there is a possibility of tuning two parameters: dipole matrix element or the mode volume. If the dipole matrix element is tuned, it is reasonable to go to the Rydberg regime. If the mode volume is tuned, it is proper to use extremely small cavities. This leaves no other option than going to the optical regime, because in μ-size cavities no microwave radiation will fit inside.

In our experiments coupling constant is made large ($g/2\pi \approx 7$ kHz) by going to the Rydberg regime. This means that the frequencies goes down automatically and the transition frequency is in the microwave regime. Accordingly, the decay constants should also be chosen. Atomic decay constant is small ($\gamma/2\pi = 0.7$ kHz) due to the Rydberg regime where the states are generally long lived, and cavity decay constant is made small ($\kappa/2\pi \approx 0.02$ kHz) by using superconductive Nb cavities with a very high Q-value.

The atoms in the micromaser play a dual role of both pumping the field and also probing the field via measurements made of the outgoing atoms. Any other means of measuring the field inside the resonator has the detrimental effect of lowering its Q-factor. The long photon storage time of the resonator allows the decay of the field to be negligible during the passage of an atom whose interaction time is much shorter than the photon storage time.

Over the last decades the closed cavity QED system in the microwave regime has been successfully used for quantum interaction between two-level Rydberg atoms and one privileged microwave mode of the radiation field theoretical and experimental studies. Since its first experimental realization

[MWM85], this system has allowed the investigation of many fundamental aspects in quantum optics.

In the recent years, various features of the quantum fields with a closed cavity systems have been investigated. In the early works devoted to the theory of the microscopic maser (Filipowicz et al. [FJM86a][FJM86b]), the case of incoherent pumping was investigated where two-level atoms, excited to their upper level, randomly interact with a single quantized mode of a superconducting cavity. When the average lifetime of a photon in the field mode is larger than the mean time between the interactions, the field mode evolves toward a steady state where some interesting non-classical features can be found. At the steady state, the radiation field inside the high-Q cavity may show highly non-classical features, like sub-Poissonian photon statistics [RSKW90] or trapping states (TS) of the cavity field [MRW88][WVHW99]. In addition to this time dependent measurements like quantum collapses and revivals of Rabi oscillations [RW87] were done. Operating the micromaser under pulsed regime and trapping conditions under the higher pump rates the generation of Fock states was also reported [VBWW00][BVW01]. These findings established the significance of the micromaser as a testing ground for fundamental principles of radiation-matter interaction.

However, first of all the observations of vacuum Rabi oscillations plays an important role in the investigation of the light–matter interactions in the micromaser. These are often used as proof of the quantum-mechanical nature of the system as they can not be explained by a classical wave theory of the radiation field. The contrast of these oscillations in real measurement can be considered as a figure of merit for the resolution of quantum features of the field. So far in the micromaser experiments at low temperatures (below 1 K) Rabi oscillations have been observed with a rather low contrast [VBWW00]. Therefore, the scope of this work is to improve the resolution of quantum field measurements with such a closed-cavity setup.

To perform such vacuum Rabi oscillation measurements, one has take care about the rate of the atoms passing through the resonator. Between the subsequent atom-field interactions, one has to wait at least 2 cavity decay cycles to be sure that the cavity is with a necessary probability in a vacuum state. Having a cavity with a Q-factor of about $5 \cdot 10^9$, a single probe experiment takes about 200 ms. For the investigation of one point with the necessary statistical depth, one needs approximately 5000 atoms. This makes the amount of time necessary for the measurement including the detection efficiency of about 50% from the given data which is about 35 min. Recording a full Rabi oscillation measurement (consisting of 10 points) takes about 6 h of pure data collection. Therefore it is important to have a reliable, stable and efficient experimental system, since any readjustments would protract the experiment. Hence, it would make it more difficult to achieve such measurements with a necessary statistical depth.

Initially due to the various shortcomings concerning improper thermal insulation and not sufficient cooling power of the ^{3}He cryostat, it was possible to run the measurements only for half an hour. The situation was further complicated by the dye-laser system that was not reliable and needed a lot of additional attention. In addition, other important experimental parts like atomic oven and detection system were working not stable enough to perform such measurements.

It turned out that almost all components of the system had to be revised either conceptually or technically. The improvements also include the refinement of several techniques that are instrumental in calibrating the system, like highly sensitive measurement of the magnetic fields. Also, the new implemented methods in the spectroscopic setup using three-step diode laser system opened the possibility of exciting other maser states that have never been done until now.

A consequence of various implemented upgrades and improvements of almost all components of the system at the end have culminated to the main result of this work. It significantly increased contrast of the quantum features of the field in resonator, vacuum Rabi oscillations, by a factor of ~ 8. This is the best result for the closed-cavity systems at the temperatures below 1 K reported so far.

Now with the resolution of quantum features as expressed by contrast of vacuum Rabi oscillations, it should be possible to attack investigations that has been beyond the scope of Garching micromaser group so far. Among the most interesting are atom-atom correlation [ELS96][CL05], quantum stochastic resonance [BM98][WSB04] or phase diffusion [CFL$^+$03] [Wal04] measurements.

The work is organized as follows: Chapter 2 discusses the main theoretical aspects of the one-atom maser physics and briefly reviews the theory of the Rabi oscillations. In Chapter 3 more detailed is described as to how the closed cavity approach has been realized in the experiment, together with the various improvements that have been performed as well as various techniques that were developed and extended to other areas. Chapter 4 describes the main experimental results. In Chapter 5, the new three-step diode laser excitation setup is presented. In Chapter 6, improved magnetic field compensation experiments are described separately. Not only a new experimental method (velocity selected magnetic field compensation) was developed but also significantly refined measuring schemes involved in realizing it (modulation-demodulation scheme). It seems that the theoretical work in connection with the development of this new method is not available outside this work (is new and recently prepared for publication). Both theory as well as the experimental realization are discussed in detail here.

Chapter 2

Theoretical Background

2.1 Basics

A single-atom maser or a micromaser allows a detailed study of the atom-field interaction. The realized situation is very close to the ideal case of a single two-level atom interacting with a single-mode quantized field.

The excited atoms are injected into a single-mode resonator at a rate low enough that at most one atom at a time is inside the resonator. In addition, it is assumed that atom-field interaction time t_{int} is much shorter than the cavity damping time κ, so that the relaxation of the resonator field mode can be ignored while an atom is inside cavity [BK86]. The micromaser is then described in the following way: while an atom flies through the cavity, the coupled field-atom system is described by the Jaynes-Cummings [JC63] hamiltonian, and during the intervals between successive atoms the evolution of the field is governed by the master equation of a harmonic oscillator interacting with a thermal bath.

The most important features are:

- Field-atom interaction is a resonant electric dipole coupling of a single mode of the field and two well-defined Rydberg states of the atom.

- Micromaser operates in the strong coupling regime:

Constant		Value (kHz)
Atom-Field Coupling	$g/2\pi$	~ 7
Atomic Decay	$\gamma/2\pi$	0.7
Cavity Decay	$\kappa/2\pi$	~ 0.02

- Field interacts at most with one atom at a time, and the duration τ of this interaction can be controlled.

To describe the dynamics that goes on in the micromaser, the description of a classical fields in metallic cavities should be looked into first. To avoid cluttering the equations with too many constants, the theory is outlined in CGS units.

A (perfect) evacuated metallic cavity is a region $\mathcal{R}$ in space with finite extension and a perfectly conducting boundary $\partial\mathcal{R}$. In such metallic cavity, a classical field is governed by Maxwell's equations with metallic boundary conditions:

$$\nabla \cdot \mathbf{E} = 0 \tag{2.1}$$

$$\nabla \cdot \mathbf{H} = 0 \tag{2.2}$$

$$\nabla \times \mathbf{H} + \frac{1}{c}\dot{\mathbf{E}} = 0 \tag{2.3}$$

$$\nabla \times \mathbf{E} - \frac{1}{c}\dot{\mathbf{H}} = 0 \tag{2.4}$$

where the boundary conditions is mathematical expression for the highly conducting wall of the cavity: $\mathbf{E} \perp \partial\mathcal{R}$. This is just a pure classical result, a

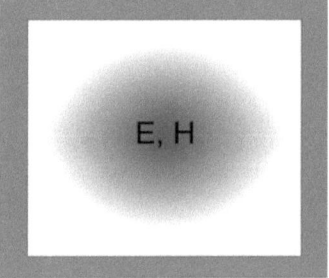

Figure 2.1: Classical field in a metallic cavity.

coupled system of partial differential equations with a boundary condition. The theory of partial differential equations tells that under mild regularity conditions, these equations for a given region and given boundary conditions can be solved where any field that is described by Maxwell's equations can be extended by splitting time $(q_k(t))$ and space $(\mathbf{A_k})$ dependence. There is a complete (orthonormal) system $\{\mathbf{A}_k\}_{k\in\mathbb{N}}$ of mode functions, i.e. divergence free metallic solutions of the vector-Helmholtz equation

$$\triangle \mathbf{A_k} = -\frac{\Omega_k^2}{c^2}\mathbf{A_k} \tag{2.5}$$

2.1. BASICS

and any field can be decomposed in the following way:

$$\mathbf{E}(\mathbf{r},t) = -\frac{1}{c}\sum_k \mathbf{A}_k(\mathbf{r})\dot{q}_k(t) \qquad (2.6)$$

and

$$\mathbf{H}(\mathbf{r},t) = \sum_k \nabla \times \mathbf{A}_k(\mathbf{r})q_k(t) \qquad (2.7)$$

Now when it comes to finding a model for a quantum field, a classical energy expression is the starting point. The energy can be represented as a sum of the energies of classical harmonic oscillators. Using the mode decomposition, the energy of any field by Jean's energy representation for classical field is

$$\mathcal{H}(t) = \frac{1}{8\pi}\int_{\mathcal{R}} d^3r \left[E^2(\mathbf{r},t) + H^2(\mathbf{r},t) \right] = \sum_k \frac{1}{2}\left[4\pi c^2 p_k^2(t) + \frac{\Omega_k^2}{4\pi c^2}q_k^2(t) \right] \qquad (2.8)$$

where p_k is the momentum canonically conjugated to q_k:

$$p_k = \frac{\partial \mathcal{H}}{\partial \dot{q}_k} = \frac{1}{4\pi c^2}\dot{q}_k \qquad (2.9)$$

Quantizing a set of independent harmonic oscillators is done by introducing the annihilation operator

$$\hat{a} \stackrel{\text{def}}{=} \sqrt{\frac{\Omega_k}{8\pi\hbar c^2}}\hat{q}_k + i\sqrt{\frac{2\pi c^2}{\hbar \Omega_k}}\hat{p}_k \qquad (2.10)$$

and dropping an infinite constant (renormalization) the result reads the quantum field description by the Hamilton operator:

$$\hat{H} = \sum_k \hbar \Omega_k \hat{a}_k^\dagger \hat{a}_k \qquad (2.11)$$

Before going further, the main micromaser parameters which are presented in the Table 2.1 are introduced. The main principle of micromaser experiment is shown in Figure 2.2. It can be described in the following steps:

- Field-atom interaction involves only one mode of the field and two atomic states: cavity geometry ensures that there is one mode (frequency Ω) that facilitates a near resonant strong coupling of two atomic states $|e\rangle$ and $|g\rangle$ (transition frequency $\omega \sim \Omega$, while radiative transitions of $|e\rangle$ and $|g\rangle$ induced by other modes are negligible.

- Incoherent pumping: atoms are prepared in energetically higher state $|e\rangle$ before they enter the cavity.

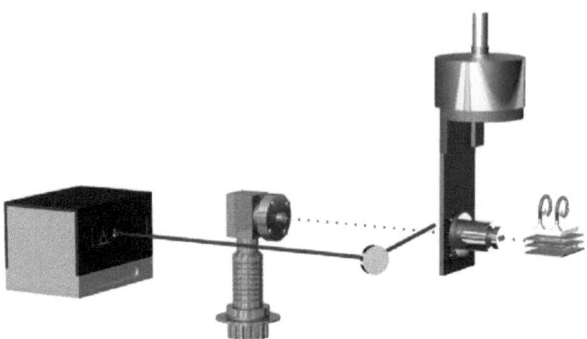

Figure 2.2: The main principle of the micromaser. The thermal atomic beam of Rb is produced in the atomic oven. With the help of laser radiation, the atoms are promoted to the upper Rydberg state $|e\rangle$ and through the small coupling holes are injected into the superconductive cylindrical cavity. The cavity is designed in a way that excited atoms can interact with one mode of the cavity field. This mode mediates a transition from upper Rydberg level to lower one. Afterwards, the state of the atoms is detected by using the state-selective field-ionization detection system. In this way the information about the field inside the cavity and its dynamics is inferred. Experiments are done with a dilute atomic beam to ensure that only one atom at a time interacts with the field. The interaction time between the field and atom is controlled using angular excitation of the laser beam by means of Doppler effect.

- Field interacts at most with one atom at a time: dilute atomic beam.

- Duration of field-atom interaction τ controlled by Doppler selection (angular excitation).

- The coupling between field and atom is much stronger than the coupling between field and environment: cavity has a very high Q-value and a very low temperature T.

Up to this point, the field was described without any coupling to the environment. Such field would stay there indefinitely. There will be no loss of energy, because the system is stationary. In the micromaser, however, the field exchanges the energy with the atoms and environment (see Figure 2.3). This is a theoretical model of basic processes in the micromaser cavity. The corresponding Hamiltonians for an atom, field and environment (thermal

2.1. BASICS

Table 2.1: Main micromaser parameters

Upper maser state	$|e\rangle$	$63P_{3/2}$
Lower maser state	$|g\rangle$	$61D_{5/2}$
Resonance frequency	$\frac{\omega}{2\pi} = \frac{\Omega}{2\pi}$	21.456 GHz
Q-factor	$Q = \Omega/\gamma = \Omega\tau_c$	$1 \cdot 10^9 \ldots 4 \cdot 10^{10}$
Interaction time	τ	$40 \ldots 120\,\mu s$
Temperature	T	~ 0.4 K

bath) are:

$$H_A = \frac{1}{2}\hbar\omega\sigma_z \tag{2.12}$$

$$H_F = \hbar\Omega a^\dagger a \tag{2.13}$$

$$H_B = \sum_k \hbar\omega_k b_k^\dagger b_k \tag{2.14}$$

and corresponding simple model for the energy exchange between an field and atom (V) and between filed and environment (W) is given by the following interaction (creation - annihilation logic):

$$V = \hbar\left(g\sigma_+ a + g^*\sigma_- a^\dagger\right) \tag{2.15}$$

$$W = \hbar\sum_k \left(g_k a^\dagger b_k + g_k^* a b_k^\dagger\right) \tag{2.16}$$

where σ_z, $\sigma_\pm$ are the Pauli-spin matrices for the atomic two-level system.

If both interactions occur at the same time, it would be complicated to describe the system. Therefore the simplification is introduced where approximate treatment of the field dynamics, i. e. of the temporal evolution of the density matrix of the field in the micromaser is following:

- While an atom is in the cavity, the weak interaction W of the field with the environment is neglected and the processes are described using Jaynes-Cummings model for field-atom dynamics;

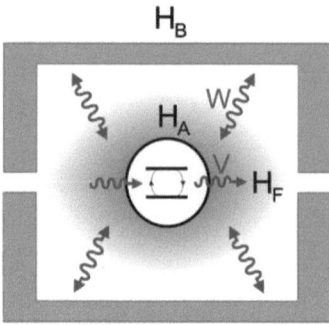

Figure 2.3: The theoretical model of micromaser: the field exchanges energy with the atoms and environment.

- While no atom is in the cavity, the interaction V is absent and the effect of the environment on the field is treated perturbatively by Master equation for the field-density operator.

Such an approach is justified because the micromaser works in the strong coupling regime. To understand the full dynamics in the micromaser the Jaynes-Cummings model and the Master equation will be discussed separately.

2.2 Jaynes-Cummings Model

The Jaynes-Cummings model is applicable if there is one mode (frequency Ω, mode function **A**) that facilitates a near resonant strong coupling of two atomic states $|e\rangle$ and $|g\rangle$ (transition frequency $\omega \sim \Omega$), while radiative transitions of $|e\rangle$ and $|g\rangle$ induced by other modes are negligible. Here ($|e\rangle$ is upper maser state and $|g\rangle$ - lower maser state.

Starting with the quantum model for the dynamics of an atom interacting with a classical field and replacing the classical field variables by their quantum analogues yields (to a very good approximation - rotating wave approximation):

$$H = H_A + H_F + V = \frac{1}{2}\hbar\omega\sigma_z + \hbar\Omega a^\dagger a + \hbar\left(g\sigma_+ a + g^*\sigma_- a^\dagger\right) \quad (2.17)$$

and the coupling constant g (in CGS units) is given by

$$g = -\frac{i}{\sqrt{2}\hbar}\frac{\hbar\omega}{\sqrt{\hbar\Omega}}\mathbf{A} \cdot \langle e| - e\mathbf{r}|g\rangle \quad (2.18)$$

2.3. MASTER EQUATION

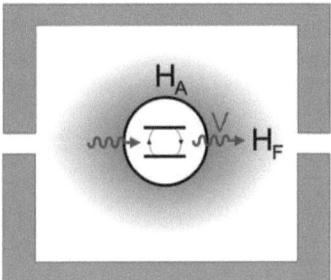

Figure 2.4: Jaynes-Cummings model for micromaser: an atom-field interaction.

Analytic solution for the field is found in the following way. Let ρ be the interaction picture density matrix of the field. If at time t the field is in a state ρ_t and an atom in the upper state $|e\rangle$ is injected and interacts for a time τ with the field then the field statistics at time $t+\tau$ is given by:

$$\rho_{t+\tau} = F_\tau \rho_t = \left[\cos(\phi\tau) - i\frac{\delta}{2}\frac{\sin(\phi\tau)}{\phi}\right]\rho_t\left[\cos^2(\phi\tau) + i\frac{\delta}{2}\frac{\sin(\phi\tau)}{\phi}\right] \\ + |g|^2 a^\dagger \frac{\sin(\phi\tau)}{\phi}\rho_t\frac{\sin(\phi\tau)}{\phi}a \quad (2.19)$$

where ϕ is the operator:

$$\phi = \sqrt{\frac{\delta^2}{4} + |g|^2(a^\dagger a + 1)} \quad (2.20)$$

The result for the atoms: The probability P_e of measuring the atom after the interaction time τ in the upper state is given by:

$$P_e = \sum_n \frac{\frac{\delta^2}{4} + |g|^2(n+1)\cos^2\left(\sqrt{\frac{\delta^2}{4} + |g|^2(n+1)}\,\tau\right)}{\frac{\delta^2}{4} + |g|^2(n+1)}\rho_{nn} \quad (2.21)$$

For a Fock state $\rho = |n\rangle\langle n|$, P_e exhibits Rabi oscillations. For detuning $\delta = 0$:

$$P_e = \cos^2\left(|g|\sqrt{n+1}\,\tau\right) \quad (2.22)$$

2.3 Master Equation

For the theoretical treatment of the dynamics of the photon field in the micromaser, a standard master equation approach is employed. The master

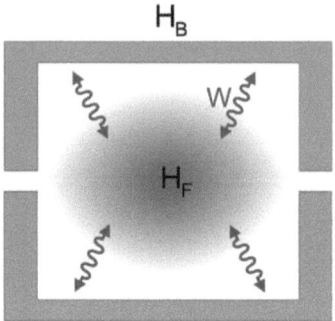

Figure 2.5: Relaxation processes in the micromaser: a field-bath interaction.

equation gives an approximate equation of motion for the interaction picture density matrix of the field ρ.

The basic assumption is done that the bath has negligible correlation time (Markoff approximation) and remains in a thermal state corresponding to a given temperature T. The changes that occur in the field on the short time scale, when the atom-field interaction happens, are ignored. In essence, the main interest are the changes that occur in the photon state during the passage of atoms.

The derivation is based on the von-Neumann equation for the full dynamics in the interaction picture and an initial product state, successive approximation to second order in W, partial tracing, coarse graining and continuous mode density approximation for the bath that yields:

$$\dot{\rho}_t = L\rho_t = -\frac{\gamma}{2}\left(n_{\text{th}} + 1\right)\left(a^\dagger a \rho_t - a\rho_t a^\dagger\right) - \frac{\gamma}{2}n_{\text{th}}\left(\rho_t a a^\dagger - a^\dagger \rho_t a\right) + adj. \tag{2.23}$$

where n_{th} is the mean thermal population number of the bath at frequency Ω:

$$n_{\text{th}} = \frac{1}{e^{\frac{\hbar\Omega}{k_B T}} - 1} \tag{2.24}$$

and the relaxation constant γ can be related to the Q-value of the cavity:

$$\gamma = \frac{\Omega}{Q} \tag{2.25}$$

2.4 Steady State

In describing the continuous micromaser operation, both relaxation and pumping processes has to be taken into account.

Let velocity selected atoms from an atomic beam fly through the cavity. Let ρ_k be the interaction-picture state of the field when the k-th atom enters the cavity (time t_k). Combining Jaynes-Cummings dynamics and subsequent relaxation, the state ρ_{k+1} is given by:

$$\rho_{k+1} = \exp\left\{-L\left(t_{k+1} - t_k - \tau\right)\right\} F_\tau \rho_k \qquad (2.26)$$

Atomic arrivals have a Poissonian distribution. Let us consider the regime where the rate of atoms be r and $1/r \gg \tau$. Then the expectation values $\langle \rho_{k+1} \rangle$, $\langle \rho_{k+1} \rangle$ are related by:

$$(1 - F_\tau)\langle \rho_{k+1} \rangle = L \langle \rho_k \rangle \qquad (2.27)$$

The thermal field state of temperature T is a steady state of the master equation. The requirement $\rho_{k+1} = \rho_k$ yields the diagonal steady state. For zero detuning, its populations are given by [FJM86b]:

$$p_n = p_0 \prod_{k=1}^n \frac{k n_{\text{th}} + N_{\text{ex}} \sin^2\left(\sqrt{k}\,|g|\,\tau\right)}{k\left(n_{\text{th}} + 1\right)}, \qquad (2.28)$$

where $N_{\text{ex}} = r/\gamma = r\tau_{\text{c}}$ is the number of atoms entering the cavity per cavity decay time $\tau_{\text{c}} = 1/\gamma = Q/\Omega$.

In the steady of one atom maser, pumped by resonant excited atoms, the energy balance implies the equality between the gain and loss rates.

In the case of an unpumped resonator, when there are no atoms, the steady state turns into the thermal state.

2.5 Vacuum Rabi Oscillations

The Rabi oscillation is the cyclic behavior of a two-state quantum system in the presence of an oscillatory driving field. A two-state system has two possible states. If they are not degenerate energy levels, the system can become "excited" when it absorbs a quantum of energy. Rabi oscillations can be interpreted as a periodic change between absorption and stimulated emission of photons.

In the micromaser, a vacuum Rabi oscillation is a damped oscillation of an initially excited atom coupled to a cavity in which the atom alternately emits photon(s) into a single-mode electromagnetic cavity and reabsorbs them.

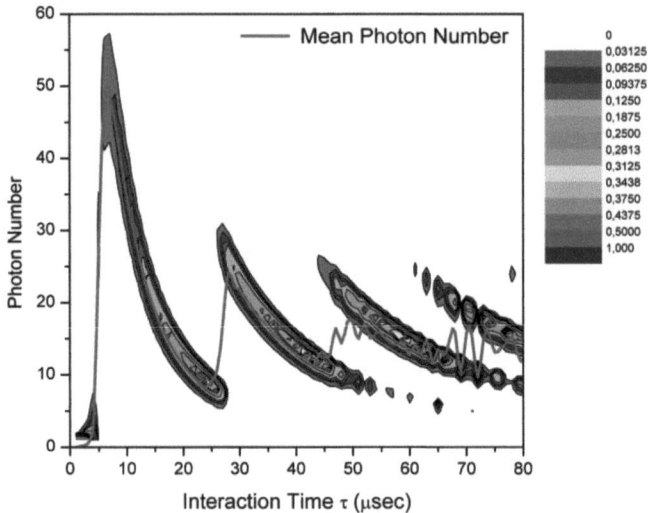

Figure 2.6: Steady state photon distribution in the micromaser.

As already shown in the Section 2.2, in the Jaynes-Cummings Hamiltonian, describing the single-atom - single-mode system, if there are no dissipative losses, therefore spontaneous emission is reversible and two level atoms undergo Rabi oscillations in the presence of a photon number $|n\rangle$, where $n = 0, 1, 2, \ldots$. Under this influence, the relative populations of the excited and ground states of an atom will oscillate at a frequency $g\sqrt{n+1}$, where g is atom-field coupling constant. Experimentally, the atomic inversion is measured, which is given by $I = P_g - P_e$ where P_g and P_e are the probabilities of finding ground and excited state atoms, respectively. For an n photon Fock state this produces Rabi oscillations which are given by

$$I(\tau) = 1 - 2P_e = -\cos\left(2|g|\sqrt{n+1}\,\tau\right) \qquad (2.29)$$

where τ is the interaction time of the atoms with the cavity field. For a vacuum Rabi oscillations ($n = 0$), as it is of the main interest in this work, the theoretical curves are shown in Figure 2.7. For the calculations, the experimental parameters were chosen, like atom-field coupling constant $g/2\pi = 7$ kHz and corresponding τ.

In the case of 0 K temperature (thermal photon number $N_{th} = 0$ also),

2.5. VACUUM RABI OSCILLATIONS

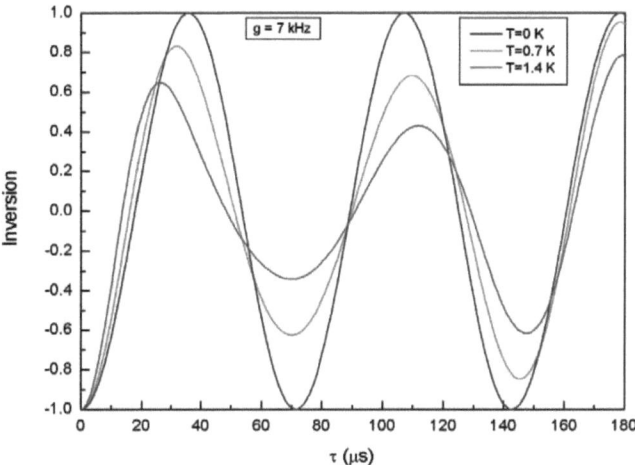

Figure 2.7: Theoretical vacuum Rabi oscillations for different temperatures, calculated using experimental parameters: atom-field coupling constant $g/2\pi = 7$ kHz and corresponding interaction time τ.

the system undergoes periodic and uniform Rabi oscillations with a maximum inversion. However in the case of the present thermal field ($T = 0.7$ K or $T = 1.4$ K correspond to $N_{th} \approx 0.3$ and $N_{th} \approx 0.9$) the vacuum Rabi oscillation periodicity gets spoiled and the contrast decreases.

The measurements of vacuum Rabi oscillations in the micromaser were done at $T = 0.7$ K temperature. The experimental setup and results are presented in Section 4.4.

Chapter 3

Experimental Setup

3.1 Overview

In the realization of the experiments, described in the theory section, atoms interacting resonantly with a single mode electromagnetic field are needed. This is achieved by using highly excited Rydberg atoms [LP79] which interact with a single mode of a high quality superconducting niobium resonator [Jac99][Kle89].

The actual realization of atomic beam generation, excitation, interaction with the cavity field and detection takes place in a cryogenic high vacuum system. This has a modular design shown in Figure 3.1. The use of a vacuum system is inevitable in cryogenic setup. Otherwise, collisions with background gas particles would preclude the formation of an atomic beam. The system consists of three sections: atomic oven chamber, auxiliary chamber and a ^{3}He cryostat. For maintenance purposes, each of the chambers can be separated from the other ones. Each chamber is equipped with its own turbomolecular pump (where these being connected to the mechanical pump system). Usually the vacuum of the order of $5 \cdot 10^{-7}$ mbar is achieved. Integrated appropriate valve system selectively allows the opening of one chamber without interfering with the vacuum in the other chambers.

In choosing the suitable element for light-matter interaction in micromaser experiments, several points should be considered:

- First, a strong dipole moment is needed to achieve a strong coupling as far as selection of the atom is concerned. Strong dipole moments can be found in the Rydberg regime where the wave functions overlap strongly and far away from the nucleus.

- Second, the transition between the states should be in the microwave regime.

- Finally, the excitation of the atoms to the Rydberg regime should be done with reasonable technical efforts. Since for the population of the

24   CHAPTER 3. EXPERIMENTAL SETUP

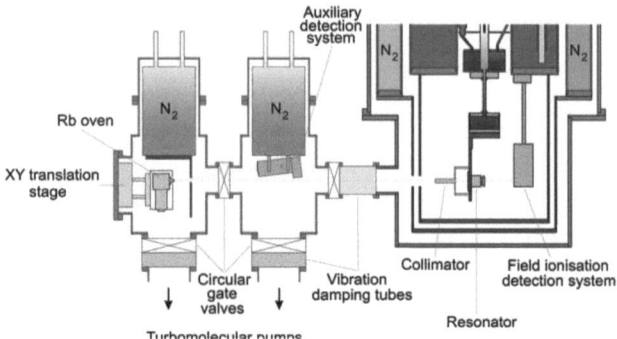

Figure 3.1: Main experimental setup: three cryogenic-vacuum chambers. Thermal beam of Rb atoms is produced in the atomic oven. Part of the atomic beam passes through auxiliary detection system, where laser frequency stabilization takes place. Main atomic beam goes to ^{3}He cryostat, where atoms interacts with the single mode resonator and afterwards are detected by field-ionization detection system.

Rydberg states the conventional lasers are used, alkali metals of higher periods are chosen.

Therefore, traditionally, rubidium (Rb) is among the most prominent members used that satisfies these conditions. The handling also plays an important role where Rb is a suitable candidate concerning low melting point and relative easy handling at room temperature.

Rubidium is a soft, silvery-white metallic element of the alkali metal group. It has a melting point at about 312 K [Lid90]. Naturally Rb is a mixture of the two isotopes: ^{85}Rb (72.2 %) and ^{87}Rb (27.8 %) [Lid90], but our experiment is tuned to operate with only ^{85}Rb isotope for quantitative reasons.

The atomic beam of Rb atoms is produced in the first section of the apparatus. With the help of laser radiation the atoms are promoted to the upper ($63P_{3/2}$) Rydberg state. They enter into the superconductive cylindrical resonator through the small coupling holes. Although the resonator has infinitely many modes, only one of them fulfills two conditions that are necessary for a significant interaction with the atom:

1. The mode frequency has to be in resonance with an atomic transition;

2. The Q-factor for this frequency should be large enough to fulfill the strong coupling condition (see Chapter 2).

For the purpose of this study, the resonator is designed in a way that only this single mode mediates a strong resonant coupling from the upper Rydberg state to the lower one ($61D_{5/2}$). Since the experiment is concerned with only very small fields (only of an order of a few quanta), it makes it very difficult to use conventional systems to access the field strength. Second, any system that allows to extract the information in a conventional way does it by extracting the intensity (energy) from the resonator. Since the experiment deals with the high Q-environment, such a method would present a leakage and significantly reduce the Q-factor. Therefore, realizing a strong coupling system from which one can extract atoms already presents a reasonable detector. Hence an indirect approach via the measuring the state of the atom exiting the resonator is used. This is done with a state-selective field-ionization system based on the principle design described in [Bab89] [MS92]. Through this, the information about the diagonal elements of the field's density matrix can be extracted by making measurements with different velocities of atoms. A complete state tomography [Ant99] [Mar03] becomes feasible if additional coherent states are mixed with the field in the resonator. For this purpose, the system is already equipped with the necessary features: the microwave coupler, which allows to send the coherent state from the microwave generator into the resonator; the whole experimental system is placed on the vibration-isolated optical table to avoid mechanical vibrations when phase-sensitive measurements are done.

This approach requires a system that deals with a significantly reduced noise levels respect to what has been reached so far. Since the figure of merit for system capability in performing such a measurements is the contrast of Rabi oscillations, this work is mainly dedicated on it.

In the following sections, the auxiliary processes (atomic beam generation, excitation and detection) for the realization of micromaser experiments will be discussed. In the last two sections of this chapter a high Q-factor resonator will be described and the cryogenic system will be explained.

3.2 Atomic Beam Production

For the micromaser operation and corresponding measurements, a stable atomic beam is necessary that provides a velocity profile with a certain width.

The atomic oven is shown in Figure 3.2. This new setup of the oven was developed to overcome the old problems which were present in previous models: such as bursts in the atomic beam (not stable operation), nozzle blocking, difficult and inconvenient control of the velocity distribution and the flux rate.

The new setup of the atomic oven consists of a two separate stainless steel cylinders (section 1 and section 2) connected by an insulating material. In the lower cylinder, a refillable cartridge of rubidium is inserted from below,

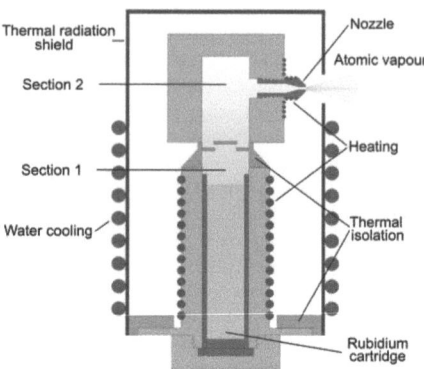

Figure 3.2: Atomic oven provides a thermal beam of rubidium atoms.

along the rotational axis. This part is heated with the resistance wire well above the Rb melting point to about 470 K temperature. The rubidium vapor emanates from the upper cylinder through a small nozzle.

Such a setup provides more flexible options in choosing the proper temperatures T_1 for section 1 and independently T_2 for section 2. This allows to access two different parameters of the atomic beam independently: overall atomic flux and the temperature profile. The temperature of the body T_1 determines how much bulk Rb is evaporated or the pressure of the oven and the tip temperature T_2 via thermalization determines the velocity distribution. This new design of the two cylinder setup is also more effective at smothering bursts of the Rb in the evaporation process.

In addition, the properties of the atomic beam can be selected using different nozzles. Longer nozzles with a smaller diameter lead to a well collimated beam with a small divergence at the expense of the atomic flux. Furthermore, since only a tiny fraction of emitted gas from the atomic oven is used in forming the atomic beam, the diameter of the nozzle should be correspondingly small to avoid vacuum contamination. For the experiments empirically a 3 mm long and 0.3 mm diameter nozzle is chosen that provides a reasonable compromise between atomic flux and divergence.

Previously, the nozzle was frequently blocked, so it turned out that is reasonable to use additional heating. For this purpose, the nozzle is heated separately to approximately 100 K higher temperature than the oven with a resistance wire. This separate heating allows also to control the flux rate

3.2. ATOMIC BEAM PRODUCTION

and velocity distribution of atoms. For the temperature monitoring a two thermocouples (type K: Chromel (Ni-Cr alloy) / Alumel (Ni-Al alloy)) are placed on the oven and nozzle.

Since the probability of an atom to leave the oven is proportional to its velocity, the velocity distribution of the atomic beam is described by a modified Maxwell-Boltzmann distribution law [Ram85, Mar03]:

$$P(v) \propto P_{MB}(v)v = \frac{1}{2}\left(\frac{m}{kT}\right)^2 v^3 e^{-\frac{mv^2}{2kT}} dv \qquad (3.1)$$

The most probable velocity for atoms is: $\sqrt{3k_B T/m}$, which in our case is about 400 m/s. The ^{85}Rb atomic velocity distribution dependence on the temperature is shown in Figure 3.3.

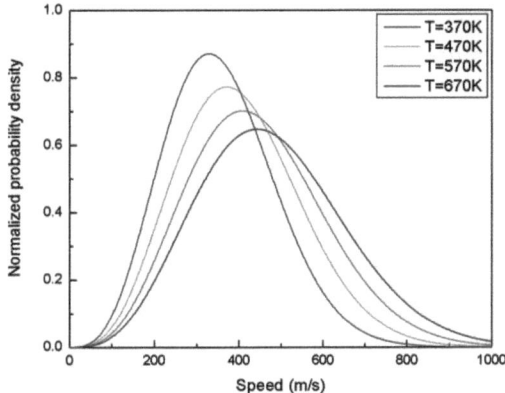

Figure 3.3: The speed probability density functions of the speeds of ^{85}Rb atoms at various temperatures.

Since the experiment needs high atomic flux of the velocities between at least 150 to 850 m/s (for e.g. Rabi oscillation measurements), it would be good to control the atomic flux and the velocity distribution independently. This was not the case before, where changing the temperature, the velocity distribution could be changed, however, at this time it would lead to the lower atomic flux in the one of the velocity distribution wings (see Figure 3.3).

As mentioned previously, a new setup of the atomic oven was made to solve this problem, where better independent control of atomic flux rate and velocity distribution of atoms are achieved.

The rubidium atoms emerging from the atomic oven are pre-collimated. The pre-collimation is done not for the purpose of forming an atomic beam, but for keeping a better vacuum in the system. The whole oven is surrounded by a water cooled copper shield, which reduces the thermal radiation. A liquid nitrogen cooled copper shield is placed in front of the oven, which traps the atoms that exits the oven in an undesirable direction. It also helps to keep the high vacuum. The fact that the large part of Rb that is not used in the atomic beam formation is condensed on this cold copper shield, it also reduces the deposition of the Rb on the other parts of the cryogenic system. This helps avoid the formation of a strong basic chemical Rb-hydroxide ($RbOH$) on these parts and the corrosion when the setup is opened and Rb reacts with the humidity.

The main experimental region, also containing the atomic beam collimator, is about 0.9 m away from the atomic oven. Concerning the dimensions of the collimator (which are explained later), the acceptance angle for the atomic beam is 0.34 deg. This means that atomic oven should be aligned to the atomic beam collimator within 5 mm accuracy. For this purpose, the whole atomic oven is placed on the XY translation stage for fine tuning and easier atomic beam alignment to the collimator in the main experimental region.

To form an atomic beam from the Rb vapor the atomic beam collimator (Figure 3.4) is used. A good collimation of atomic beam is important also for velocity selection, because this is done via the Doppler effect where a major role plays the velocity of the atom and the angle (see Section 4.3). Smaller diameter collimator would lead to smaller angular dispersion of an atomic beam, therefore the narrow velocity distribution in the excitation region, but at the same time also to smaller atomic flux. In order to guarantee a reasonable compromise between atomic flux and angular dispersion of atoms in the beam, the dimensions of the collimator were chosen correspondingly. The length of the collimation tube is 50 mm and the inner diameter is 0.3 mm, which gives an acceptance angle of 0.34 deg and therefore the velocity dispersion of $\sim$ 12 m/s ($\sim$ 1.85 MHz), which is of the order of the linewidth of the laser.

Before, collimator was made of several brass parts that caused several problems: mechanical instability leading to misalignment with respect to the atomic beam during the cooling process due to different tension of materials; contact potentials causing the atomic line shifts in the excitation region; complicated mounting and aligning with other experimental parts procedure which cost some time. In order to overcome these problems, a new atomic beam collimator together with the excitation region (Figure 3.4) was designed. To assure high accuracy these two parts (collimator and

3.3. EXCITATION

Figure 3.4: Atomic beam collimator with the excitation region manufactured from a single piece of Nb.

excitation region) were manufactured using CNC (computerized numerical control) technique and combined using electron welding technology. Since the collimator is mechanically directly attached to the resonator, it is produced from the same material (Nb) to avoid the contact potentials. It is attached using specially manufactured screws from Nb also.

3.3 Excitation

The atoms are promoted to the Rydberg regime via three-step excitation ladder using new designed diode laser system.

The selective excitation of Rydberg states requires the use of frequency selective lasers. As already discussed in the Section 3.1, in the experiments Rb is used, due to the excitation to the Rydberg regime possibility using conventional laser technologies. Therefore, traditionally, frequency doubled dye lasers are chosen for this purpose [Ant99] [Bra01].

For the one-step excitation in the experiments, an Argon-Ion (Ar^+) (Coherent Innova-300) pumped intracavity-frequency-doubled Rhodamine-6G-dye laser (Spectra-Physics 380D) was used providing the necessary radiation in the UV at 297 nm. However, there were several problems with using this laser system, mainly:

- laser beam spatial instability ($\sim$ 0.1 deg/s), leading to fluctuating excitation rates - instable micromaser operation;

- bad laser mode quality related to the crystal degradation, which caused problems with the focusing and correspondingly reduced the efficiency of excitation;

- laser power instability - due to misalignment laser power dropped $\sim$ 10 %/h, and this made an additional active intensity stabilization necessary. At the same time, this meant that not the maximum laser power was available for the experiment, but only a floor, chosen in a way that the laser would remain for a sufficient amount of time beyond this value;

- frequency doubler D-ADA crystal for which intracavity frequency doubled laser is optimized degradation (deuterated ammonium dihydrogen arsenate crystals are no more produced);

- frequent ($\sim$ every 2 h) realignments were necessary; weekly dye change should be performed.

To get rid of these problems, a new diode laser system has been constructed to realize the promotion of the atoms to the upper maser state in the following three steps:

$$5S_{1/2}, F = 3 \longrightarrow 5P_{3/2}, F = 4 \longrightarrow 5D_{5/2}, F = 5 \longrightarrow 63P_{3/2}$$

In the micromaser experiments, a few attempts have been made to set a three-step-excitation diode laser systems in the past. However, due to inefficiency in the generation of the Rydberg states these projects were abandoned.

This new three-step-excitation diode laser setup is the first one proven to be capable of replacing the conventional one-step-excitation dye lasers in the micromaser experiments. At the same time, the new setup also solved many problems that were present in the past using the common dye laser systems. In using such a diode laser system, the beam pointing stability or bad laser mode comes not in question because of the rigid resonator design. The comparison of the old dye-laser and the new developed three-step diode-laser system is reviewed in the Table 3.1.

Table 3.1: Comparison of the old and the new one laser systems

Parameter	Old dye-laser	New diode laser system
Max. count rate (1/s)	$\sim 10^4$	$\sim 10^6$
Continuous locking (min)	~ 20	~ 480

The new three-step laser setup (Figure 3.5) employs three commercially available grating-stabilized diode lasers. The main parameters of each laser stage is presented in the Table 3.2.

The first two stages are frequency stabilized on the spectroscopic signals generated in the Rb cells, and the third stage laser is locked on the atomic beam signal.

3.3. EXCITATION

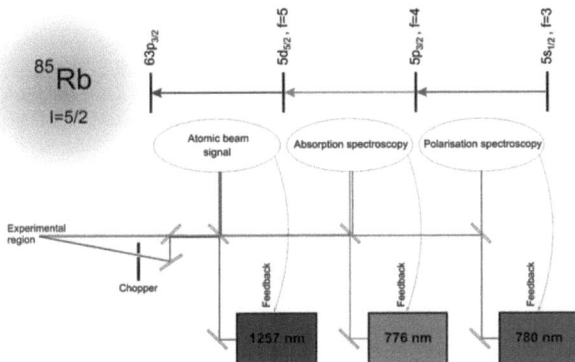

Figure 3.5: New three-step diode laser setup and ^{85}Rb atomic excitation scheme.

Table 3.2: The main parameters of the corresponding diode laser stages

Stage	Wavelength (nm)	Power (mW)
I-st (Toptica DL 100 L)	780.243	80
II-nd (Toptica DL 100)	775.978	40
III-rd (Toptica DL 100 L)	1256.730	25

The spectroscopy setup for the first stage is quite simple and straightforward. The transition here is resolved by means of the saturation spectroscopy.

For the second stage, all the approaches so far relied on an indirect method to detect the right frequency on spectroscopy of this transition via the cascaded decay [Lan94]. In the new approach, we detect the $5P_{3/2}, F = 4 \longrightarrow 5D_{5/2}, F = 5$ transition directly. In the diploma thesis of Th. Germann [Ger08] the comparison of the old and new one approaches in terms of the signal-to-noise ratio has been performed indicating that the new approach is superior by an one order of magnitude. This new method for stabilizing the second stage proved to be so simple and successful that it has been applied in the laboratory for other experiments, like a direct detection of the Rydberg states in the cells [THS+09] [TGH+09].

The frequency stabilization of the third stage is done directly on the atomic beam. For this purpose, all the three laser beams are overlapped and directed to atomic beam in the cryogenic vacuum system. The excitation to the Rydberg regime ($63P_{3/2}$) and the third stage frequency stabilization takes place in the "auxiliary" chamber, where all the three lasers hit the atomic beam perpendicularly.

This flexibility of using the three step excitation offers the possibility of populating the Rydberg states selectively. This can be done by using various atomic selection rules in order to selectively populate corresponding states for different experiments. This selective excitation will be discussed later together with the description of the corresponding experimental measurements and results.

All the lasers are frequency stabilized on the peak of the corresponding spectroscopic and atomic beam signals. This was not the case in the old setup, where the side of fringe (constant count rate) stabilization was performed what lead to undesired frequency changes. In the new setup the stabilization is done using adapted synchronous demodulation (lock-in) scheme. Error signals are processed with a specially designed home made regulators with a modified PID topology.

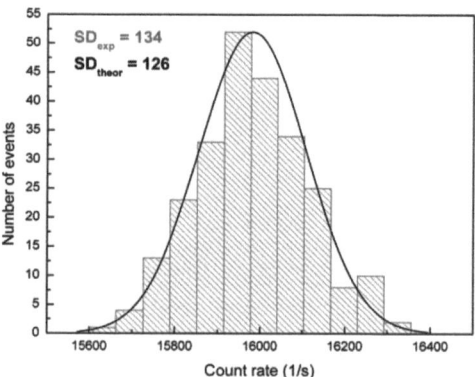

Figure 3.6: Test of the stability of the system: excited atomic beam statistics over the time should show Poissonian distribution.

How well and stable such a three-step laser system with a specially adapted stabilization units works, can be extracted from the excited atomic beam statistics. The result is shown in Figure 3.6. If all the three lasers are truly monochromatic, then the count rate (discrete signal) statistics for the perfect system should show a Poissonian distribution (discrete probability distribution). This expresses the probability of a number of events occurring in a fixed period of time, if these events occur with a known average rate and independently of the time since the last event. However, any gitter would result to a broadening of the signal. Therefore the proximity of the observed distribution to Poissonian one is a figure of merit for a quality of atomic beam generation, stability of the three lasers in terms of power and frequency and detection. Actually it is more than just a test of a laser stabilization, because it includes all main parts of the experiment. The experimental results are very close to the Poissonian distribution - theoretical limit (within 6 %), that shows not only good stabilization of the three diode laser system, but also the reliable and stable production and detection of the atomic beam.

Since some new methods concerning laser, spectroscopical and stabilization setups were realized, that have quite lengthy descriptions, it will be outlined in a separate chapter. The full description about the new rating-stabilized diode laser system, excitation realization, used schemes and corresponding optical setups will be discussed further in Chapter 5.

3.4 Detection

The information read-out about the field inside the resonator is done by detection of atoms with the state-selective field-ionization detection system. Auxiliary detection setup is used for laser frequency stabilization.

The only possibility of gaining the information about the field (and it's dynamics) inside the resonator is by observing its influence on atoms. This is done by detecting the atomic state of the atoms by means of state-selective field-ionization [TLP76] and detection of electrons in two separate single channel electron multiplier detectors (channeltrons).

A channeltron is a small, curved glass vacuum-tube structure that multiplies incident charges (at the output end a pulse of 10^7 to 10^8 electrons emerges; pulse duration is ~ 10 ns). For best performance, the Burle 7010M single channel electron multipliers in the form of a planar spiral glass tube with a 10 mm diameter input cone in this experiment are used. For the detection of electrons, the input of channeltron is kept grounded and the output is at a high positive voltage (3 kV). At such voltages, the typical minimum gain is about $5 \cdot 10^7$ [Bur]. The channeltron in the micromaser experiments is operated in pulse counting mode where produced output pulses have characteristic amplitude, whereas other possible analog operating mode of the channeltrons have a very wide distribution of output pulse amplitudes. The operating point for a channeltron in the pulse counting mode is usually determined by the point at which a plateau is reached in the count rate versus voltage characteristic curve. A typical measured curve is shown is Figure 3.7. The characteristic curve shows four regimes:

1. Low gain: at the low voltages applied potential is not enough to multiply incident charges - no pulses are produced, or the pulse amplitude

is very small, i.e. below threshold.

2. Threshold: as the voltage applied to the channel is increased, the gain rises and the output pulses become larger. The pulses are not of the same size, but as the gain increases, more of them exceed the equivalent threshold. The process continues until all the pulses are above threshold.

3. Plateau: the plateau occurs when all the signal is being collected at the input of the channeltron. Additional increase in voltage raise the gain, but the count rate remains essentially constant.

4. Ion back action: still increasing the voltage a point is reached where ion feedback (parasitic effect when gasses adsorbed on the surface of the walls are desorbed and ionized, forming positive ions, which travel back toward the input of the device and when they strike the walls near the input, they produce secondary electrons which are subsequently amplified and detected at the output end as a noise pulse) becomes significant due to very high gain and the count rate again increases rapidly. This is an undesirable condition since the extra counts are produced within the channeltron itself and are not the result of an input.

The middle point of the plateau was used as the operating point for the stability reasons. If the voltage here changes a little bit, the pulse height distribution changes a bit also, but the count rate remains the same. For this reason, the operating voltage of the channeltron is set in the middle of this regime, which is at 3050 V.

As the channeltron ages, the plateau moves to the right and voltage must be increased. It either is the result of continuous degradation effected by high velocity electron scrubbing and therefore causing the surface degradation and reduction of concentration of elements that are necessary to reach a high gain, or the contamination with Rb, that reduces the gain of the channeltron also. The lifetime of channeltrons, that is significantly below the ones specified by the suppliers, was observed. Therefore it is assumed that the contamination with the Rb causes the main problems.

For the ionization of atoms, a certain static electric field is necessary. As the principle quantum number n of an atom increases, the size of the orbit and also the coupling of the electron with the applied electric field increases as n^2, whereas the binding energy decreases as n^{-2} [ZLKK79]. The strength of ionization field depends on effective principle quantum number n^* [KLZ83]:

$$E = \frac{R^2 \epsilon_0 \pi}{n^{*4} e^3} = \frac{3.21 \cdot 10^8}{n^{*4}} \left[\frac{V}{cm} \right], \tag{3.2}$$

3.4. DETECTION

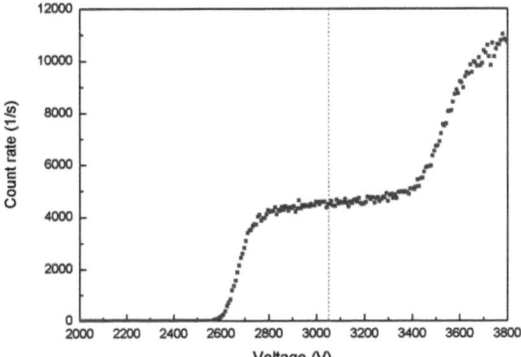

Figure 3.7: Plateau curve of the channeltron. The channeltron is operated in the middle of the plateau, in this case at 3050 V.

where R is Rydberg constant, e is elementary charge and ϵ_0 is electric constant.

For Rydberg states, the atoms are already ionized in the electric field of about several tens of volts (for $n^* = 61$ $U_I = 23.2$ V/cm; and for $n^* = 63$ $U_I = 20.4$ V/cm correspondingly).

Such field-ionization process happens in a well shielded device - channeltron box (Figure 3.8) - which works at plate condensator principle. Here the atoms are subjected to adjustable electric field created between the negative voltage electrode-plate and positively charged ionization electrode-grids. The field gradient is achieved due to intermediate specially widening-shaped and grounded plates. These plates are shaped in a way, that along the flight path of atoms the electric field slowly increases, in this way creating a different strength of electric field in two regions nearby the channeltrons, and in between remains constant for better spatial separation, i.e. for avoiding the miscounts in corresponding channeltrons. State selective ionization is achieved by detecting the electrons in two channeltrons.

Atoms in the upper maser state $63P_{3/2}$ are ionized by lower electric fields, so along the path in the channeltron box they will be ionized earlier than the atoms in the lower maser state $61D_{5/2}$ due to the field gradient described earlier. By choosing the right potential of electrodes, the situation is reached where atoms with two different states are detected in two separate channeltrons.

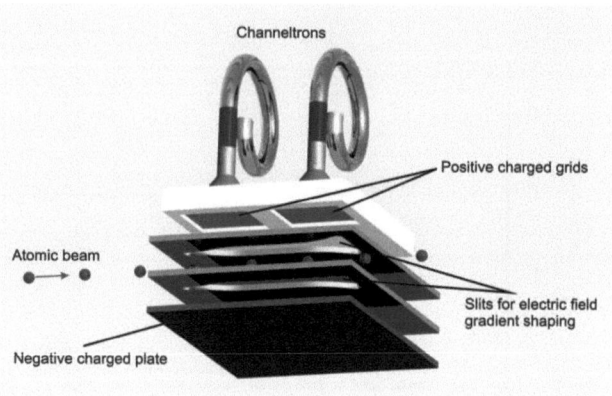

Figure 3.8: Channeltron box: state-selective field-ionization detection system. During the passage, atoms experience increasing electric field. Atoms with different atomic states are ionized at different points. Detection is done via channeltrons by counting ionized electrons.

The calibration of the working voltage for electrodes is done in the following way. Initially the resonator (see Section 3.5) is tuned out of resonance, to have all the atoms in the initially prepared upper maser state. Then the electrode voltage is scanned, the signal is detected in both of the channeltrons and recorded with a computer (Figure 3.9(a)).

The same is done with the resonator tuned in the resonance to have the highest possible count rate of maser ground state atoms (Figure 3.9(b)). In both cases, first the signal in channeltron 2 is observed, in front of which the electric field is highest due to the special geometry of slits. At a certain threshold value, the field becomes so big that the atoms are already ionized in the first channeltron and the signal rises strongly, while it disappears in channeltron 2. The ionization curves for the maser ground state atoms (measurement with resonator in resonance) are shifted in the direction of higher voltages compared to the non-resonant measurement, which reflects higher ionization-potential for the energetically lower lying state.

The optimum value of the applied ionization voltage lies where in channeltron 1 as high as possible number of atoms in the upper maser state are detected, and correspondingly as few as possible in the ground state. Analogously in channeltron 2 as high as possible number of atoms in the ground state should be observed, and as few as possible one in the upper maser state. In this case the difference between the count rates for resonant and

3.4. DETECTION

non-resonant maser for channeltron 1 should show a pronounced maximum and for channeltron 2 - pronounced minimum. Such difference of the count rates is depicted in the Figure 3.9(c). The optimum ionization voltage in this case lies at 152 V, which is indicated in the diagrams by dashed vertical line. This value is higher compared to the theoretical one discussed earlier, but here the distance between the ionization plates of 2 cm should be taken into account, and of course the field gradient forming plates in between.

In using the old channeltron box some problems were encoutered, like insufficient state selectivity in each of the channeltrons, saturation effects at low temperatures, and unstable long term operation of the whole state-selective field-ionization system.

Several improvements to the channeltron box have been done to solve these problems. To achieve a more precise calibration of the ionization voltage and better state selectivity in both channeltrons, the ionization grid voltage for each channeltron is now controlled independently. The final manual offset-ionization-voltage adjustment for each channeltron allows to get steeper ionization distribution curves and in this case better discrimination of both maser-states in the channeltrons compared to the old channeltron box setup.

Previous set up of channeltron box was also not stable in long term operation, due to the charging of the insulators, that hold the channeltrons, by free electrons. So all the channeltron box was revised to properly shield the insulators, which leads to stable long term operation.

The electrical resistance of the channeltrons at room temperature is about 600 MΩ. Due to the cooling to liquid He temperatures the resistance of the channletrons increase by more than two orders of magnitude. That leads to a channeltron saturation effect because the electrons cannot be replenished on the timescale of the pulse transit, imposing an ultimate limit on the output count rate. To avoid this the channeltrons are wrapped with special high resistivity NiCr20AlSi wire (ISAOHM®) and heated slightly above 77 K temperature.

For the laser frequency stabilization to an atomic beam (see Section 3.3) a similar auxiliary detection system is used, which works at the same principle like the one described above. This setup (Figure 3.10) for easier maintenance is placed in a separate vacuum chamber and is aligned in a way that a portion of the atomic beam by the small angle up from the atomic oven is injected. ^{85}Rb atoms passes through a 0.5 mm diameter and 50 mm long collimator and are excited to the Rydberg state ($63P_{3/2}$) with the perpendicularly crossing laser beam. At the end, atoms are detected using the same field-ionization technique, as described above. Ionization is done by applying an orthogonal static electric field to the atomic beam direction using a pair of plates. The ionization region can be kept in a quite narrow zone with the help of one tilted negative-potential ionization plate, which produces a field that is increasing into the atomic beam propagation

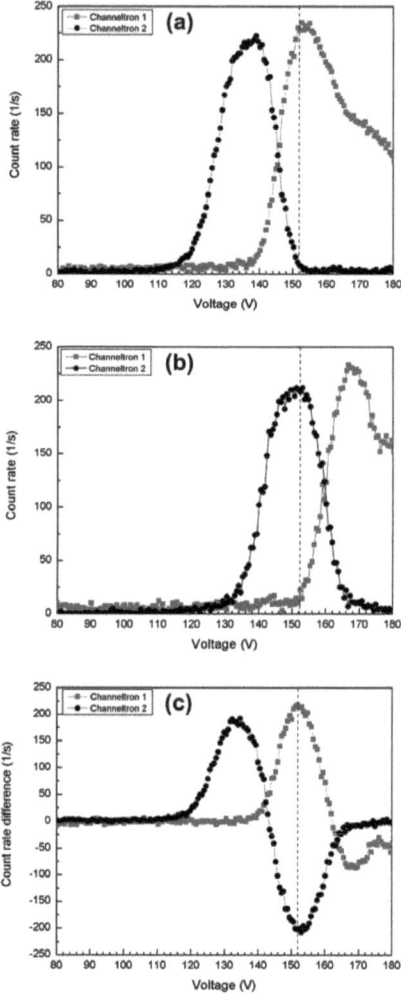

Figure 3.9: Field-ionization signal for atoms in the upper-maser state and in the lower maser state. The graphs shows the signal in both channeltrons in the case of non-resonant resonator (a) and the resonator in resonance (b). In graph (c) is depicted the difference signal of the (a) and (b) curves. The optimum ionization voltage in this case lies at 152 V.

direction. The ionization fragments - electrons - emerge through an array of small holes in the middle of the positive charged plate and are detected by single channel electron multiplier detector.

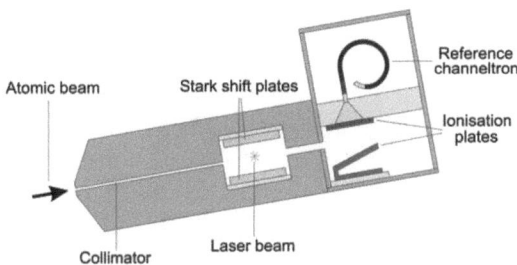

Figure 3.10: Auxiliary detection system for the laser frequency stabilization on the desired atomic transition.

Since the channeltron works in the pulsed regime, at the end there is a series of voltage pulses, the number of which are proportional to the number of ionized atoms. Using this signal as an error one, the laser is stabilized on the top of the desired ^{85}Rb $63P_{3/2}$ atomic line. The laser stabilization technique will be discussed in more detail in Section 5.2.

In the excitation region perpendicularly to the atomic and the laser beams, there are two parallel capacitor plates. Here one can apply an electric field and shift the spectral lines of ^{85}Rb due to the Stark effect. Since the laser is stabilized on the peak of the atomic transition, it changes the frequency also due to the shift. This precisely controlled laser frequency shifting is used for the velocity selection of the atoms in the main experimental region and is discussed in Section 4.3.

3.5 Resonator

The main part of the one-atom-maser experiment is a cylindrical high quality factor superconductive resonator within which the atom-field interaction takes place.

For the generation and storage of microwave photons, a superconductive cylindrical resonator is used, which is developed in Max-Planck-Institute of Quantum Optics.

The length and the inner diameter of the resonator is about 25 mm. Symmetrically in the center of both sides of the cylinder are holes of about 2 mm diameter for the entry and exit of the atomic beam, as well as for microwave injection.

Figure 3.11: The main part of one-atom-maser experiment: high Q-factor superconductive cylindrical niobium resonator.

The atomic beam passes through the cylindrical resonator along its axis, where only the TE_{1np} and TM_{1np} modes posses a non-vanishing transversal electric field. The indexes m, n and p for the classification of the TE- resp. TM-modes have the following meaning [PP71] [TS75] [Riz88]:

m: the number of half-period variations of E_r with respect to θ (of H_r for TM modes);

n: the number of half-period variations of E_θ with respect to r (of H_θ for TM modes);

p: the number of half-period variations of E_r with respect to z (of H_r for TM modes).

The TE_{121} mode is used in the experiment, while along the atomic beam axis it has uniform direction high strength of electric field. Mode form corresponds to a half period of a sin wave. The polarization direction allows a simple coupling of a microwaves through the waveguide which is important for the resonator frequency and Q-factor measurements. The resonance frequency of cylindrical resonator is given by [Jac99]:

$$\omega = \frac{1}{\sqrt{\mu\epsilon}}\sqrt{\frac{x_{12}^{'2}}{R^2} + \frac{\pi^2}{L^2}}, \qquad (3.3)$$

where μ is magnetic permeability, ϵ is the dielectric constant, R and L are the radius and length of the resonator respectively, $x_{12}^{'}$ is the 2-nd zero of the

3.5. RESONATOR

first derivative of the Bessel function J_1 ($x'_{12} = 5.331$). The mode volume of TE_{121} mode is $V = 1.58 \cdot 10^{-2} \cdot \pi R^2 L$. The distribution of the transverse E-field in the resonator is depicted in Figure 3.12.

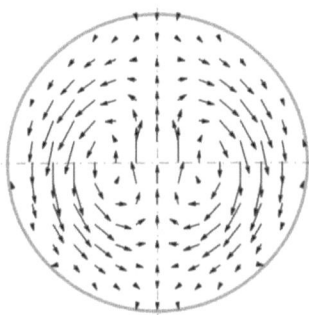

Figure 3.12: The transverse electric field distribution of the TE_{121} mode in the resonator.

In the ideal cylindrical resonator, the TE_{121} mode is doubly degenerate. The degeneracy is removed by a slight deformation of the circular cross section into an oval shape, which then determines the direction of polarization of the field mode. The deformation is achieved by squeezing the resonator (described later). Because the frequency depends inversely on the spatial dimensions, this causes one degenerate resonance to be shifted towards higher frequencies, whereas that for orthogonal polarization shifts to lower values. In this experiment the vertical polarization mode is used, which is tuned by squeezing to the higher frequencies.

Resonator is made of pure 99.9 % niobium. At atmospheric pressure the critical temperature for superconductivity of niobium is 9.3 K. Since the unloaded electrical quality factor of a vacuum superconducting cavity resonator is given by [Lui01]:

$$Q_0^{-1} = R_s \Gamma^{-1} \tag{3.4}$$

where R_s is the surface resistance and Γ is the geometric factor of the mode, to reach high quality factor one needs low R_s. To surface resistance R_s mainly contributes two factors: the residual resistance (which can be related to normal conducting inclusions, surface oxides, interface and dielectric rf losses) and so-called BCS term (due to the unpaired electrons at the Fermi level) [Mül83]. The surface resistance dependence on the temperature for the niobium resonator with the parameters used in the experiment is shown in Figure 3.13. Due to the high-advanced preparation technique of the niobium,

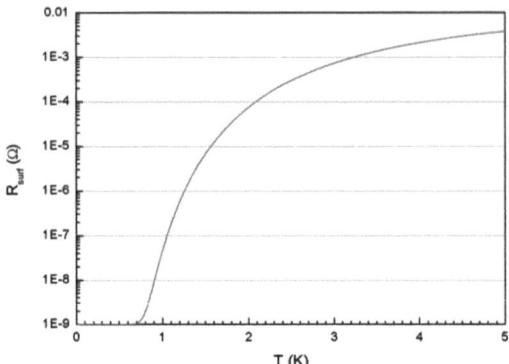

Figure 3.13: Surface resistance R_s dependence on the temperature for the 21.456 GHz niobium resonator.

surface resistance of the superconductive resonator can reach several nΩ, leading to Q-factors of the order of 10^{10} [AFH+71].

The production of such a resonator requires great precision and care and takes several steps:

- First of all the pot and the lid of resonator is made from niobium with the best possible precision using CNC-machines (computerized numerical control).

- After electro-polishing both parts are welded in a vacuum with an electron beam welding (pro-beam AG&Co.KGaA) technique.

- Then the resonator is etched (buffered chemical polishing) in the 1 : 2 : 1 mixture of the nitric (HNO_3), phosphoric (H_3PO_4) and hydrofluoric (HF) acids.

- And finally the resonator is baked in ultra high vacuum oven at about 1800 °C for 24 h.

Afterwards, the resonator is placed in the He-bath cryostat, cooled down and its properties at superconductive temperature are thoroughly tested. If the resonant frequency lies not about 15 MHz below the desired atomic transition frequency of 21.465 GHz, which can be tuned by reversible mechanical squeezing, the etching and baking steps have to be repeated.

3.5. RESONATOR

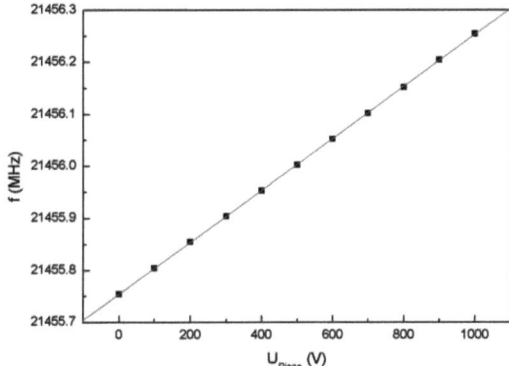

Figure 3.14: Resonator frequency dependence on the applied piezo voltage. Resonator with the mechanical screwdriver is squeezed so, so that the transition frequency (21.456 GHz) lies in the center of the piezo tuning range.

Baking procedure is also important for the high Q-factor, since it reduces lattice irregularities, builds 1 - 10 mm single crystal domains, reduces light-element concentration, homogenizes their distribution and reduces the thickness of surface oxide layer.

In maintaining high Q-factor, baking procedure should be repeated every time the main cryostat is opened and resonator is left even for short time in air.

Adjustment of the frequency to the required atomic one ($63P_{3/2} \rightarrow 61D_{5/2} = 21.456$ GHz) is performed by elastic mechanical deformation of the resonator cylinder. Rough tuning in the range of 15 MHz is done manually with mechanical screwdriver and the fine sweep in 500 kHz frequency range by 0.5 kHz steps is done with the piezoelectric drive. Figure 3.14 shows the measured resonator frequency tuning range as the applied piezo voltage is changed. The piezoelectric unit together with the resonator is placed on the coldfinger of the ^{3}He cryostat and cooled down to a temperature of several hundred mK.

The resonator Q-factor is measured by a heterodyne technique, since a direct measurement of microwave frequencies is complicated. The basic setup of a measurement is depicted in Figure 3.15.

In the first the resonance of the resonator is located by feeding the signal (with Wandel&Goltermann tracking generator TG-23) via the rectangular

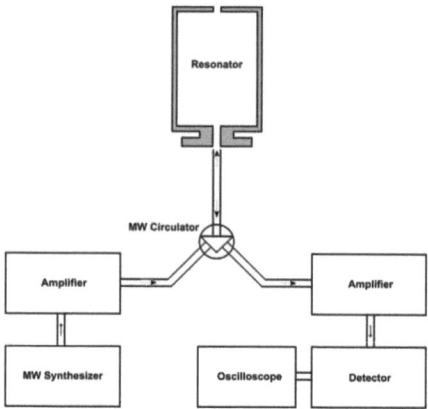

Figure 3.15: Resonator Q-factor measurement setup. A synthesizer produces a microwave which is periodically switched between two frequencies. One of them is resonant with the resonator and is coupled in. During the radiation of second frequency it mixes with the leaked out signal of the resonator. This beat signal through the microwave circulator is transmitted to the detector.

K-band microwave waveguide into resonator and observing coupled out signal with the spectrum analyzer (Wandel&Goltermann SNA-33) which comes through circulator and microwave waveguide back. Then a microwave signal generated from a synthesizer (model Systron Donner 1730B) which is periodically modulated between two values - one resonant with the resonator and the other detuned by 100 Hz is send via microwave waveguide to resonator. Inside the resonator, the resonant wave creates a field which decays exponentially. A part of this field is transmitted through the coupling holes out of the resonator. The reflected from the resonator detuned microwave signal superimposes with the coupled out resonant signal and propagates back through the rectangular waveguide and microwave circulator where finally is detected in microwave diode. The electric field is:

$$E(t) = E_0 e^{-i\omega_0 t + \phi_0} e^{-\gamma t} + E_1 e^{-i\omega_1 t + \phi_1} \tag{3.5}$$

and the intensity which is detected by the microwave diode is proportional to:

$$|E(t)|^2 = |E_0|^2 e^{-2\gamma t} + |E_1|^2 + 2E_0 E_1 \cos\left[(\omega_0 - \omega_1)t + \phi_0 - \phi_1\right] e^{-\gamma t} \tag{3.6}$$

Since $E_0 \ll E_1$, the quadratic term of E_0 can be neglected. So detected

3.6. ³HE CRYOSTAT

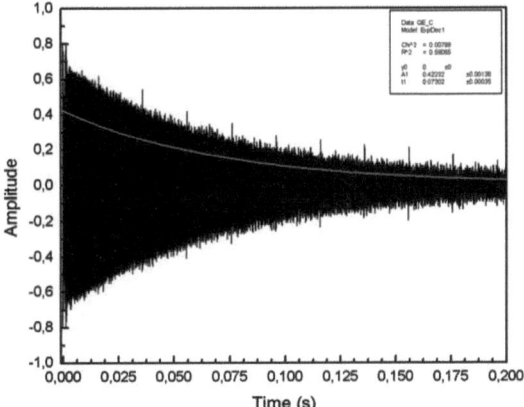

Figure 3.16: Resonator Q-factor measurement. From a fitted red curve extracted energy decay time is 73 ms leading to Q-factor value of $4.9 \cdot 10^9$.

intensity can be written as:

$$|E(t)|^2 = |E_1|^2 + 2E_0 E_1 \cos\left[(\omega_0 - \omega_1)t + \phi_0 - \phi_1\right]e^{-\gamma t} \quad (3.7)$$

where $|E_1|^2$ term is filtered out by high pass filter. This shows that the beat signal also decays with the time constant of the electric field. From the decay time of the amplitude of the beat signal one can directly extract the photon lifetime and the resonator Q-factor:

$$Q = \pi \nu_{res} \tau_f, \quad (3.8)$$

where τ_f is resonator field decay time and ν_{res} the resonator frequency.

A typical result of such a heterodyne measurement is shown in Figure 3.16. Corresponding to extracted from the fit a decay time of 73 ms the resonator Q-factor is $Q = 4.9 \cdot 10^9$.

For a newly produced resonator, Q-factor values up to one order of magnitude higher is achieved [Wal04]. However, this is the subject of fluctuations and depends on individual manufacturing process.

3.6 ³He cryostat

The cryostat is used to make experiments at very low temperatures for sup-

46 CHAPTER 3. EXPERIMENTAL SETUP

pression of the thermal radiation and achieving high Q-factor values of the cavity.

The one-atom-maser experiments should be performed at very low temperatures for two reasons:

- First of all, the thermal photon number in the cavity should be reduced in order to investigate only pure quantum atom-field interaction (see Chapter 2). Figure 3.17 shows the thermal photon number n_{th} dependence on the temperature. Often an important requirement in either theoretical or experimental consideration is that there is no thermal photon ($n_{th} < 0.5$) in the cavity. This requires that the micromaser cavity is cooled down to below 0.9 K.

- Second, the superconductor surface resistance, which plays a significant role for the quality factor of the cavity, is inverse proportional to its temperature (for details see Section 3.5). So in achieving high cavity Q-factors the temperatures should be below 0.7 K (see Figure 3.13).

^{3}He cryostat is used for these purposes in this experiment.

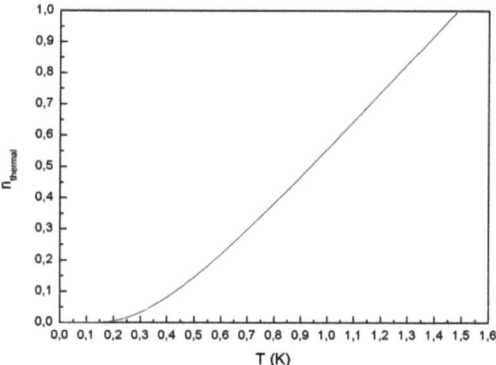

Figure 3.17: Mean thermal photon number dependence on the temperature.

One of the simplest ways for reaching temperatures significantly below 1 K is by using a ^{3}He evaporation cryostat. Whereas by pumping on a liquid bath of ^{4}He a temperature of approximately of 1 K can be obtained routinely, the lower limit with ^{3}He is slightly below 0.3 K. It is because of

3.6. ^3HE CRYOSTAT

the smaller mass of the ^{3}He atom, therefore larger zero point motion than ^{4}He. The vapor pressure of ^{3}He is also at all temperatures higher than the vapor pressure of ^{4}He. The normal boiling point of ^{3}He is 3.19 K, its critical temperature and pressure are 3.32 K and 116 kPa respectively, and the latent heat of evaporation at 0.3 K is 21 J/mol [Wil67].

It is also worth mentioning that the interval from 3.3 K to 0.3 K which can be covered by pumping on liquid ^{3}He, is more than an order of magnitude in temperature.

A schematic drawing of a modified commercial Oxford ^{3}He cryostat is shown in Figure 3.18. Nitrogen and helium baths, under standard atmospheric pressure at 77 K and 4.2 K respectively, surround a vacuum chamber inside of which below the ^{3}He pot on the cooper plate (so called coldfinger) main experimental equipment - cavity with the microwave coupler, squeezing mechanism and collimator - is mounted. The 1 K pot is equipped with a narrow (about 1 mm diameter) tube connected to ^{4}He bath. The inflow of liquid ^{4}He from the 4.2 K bath is controlled precisely with the needle valve. ^{4}He is pumped away from the other side of the 1 K pot by means of a mechanical vacuum pump, cooling it down to near 1 K. This 1 K pot is used as the heat sink: all tubes and electrical wires going to the ^{3}He pot and coldfinger are thermally anchored to it. At this temperature also a ^{3}He gas from the dumps (reservoir for storing ^{3}He in vapor-form) starts to condense in the ^{3}He pot. Finally, when the ^{3}He pot becomes filled with liquid by pumping on it a base temperature of 0.35 K can be achieved. The pumping on ^{3}He is done with the passive charcoal sorption pump, which starts to absorb gas when cooled down below 40 K. The pump is brought into operation by lowering its temperature by heat exchange with the ^{4}He bath. The flow of ^{4}He through the heat exchanger is increased by a small vacuum pump and the rate of flow is controlled by a valve in the pumping line. A heater is fitted to the sorption pump so that its temperature can be controlled. Due to the high costs of ^{3}He, it is handled in a closed system.

This ^{3}He cryostat is of the single-cycle type. Once the ^{3}He pot is filled, low temperatures can be maintained as long as there is liquid in the pot. On the other hand, the sorption pump has also its capacity limits: when it becomes saturated it has to be warmed up to regenerate the absorbed material (^{3}He).

In the experiments at these low temperatures, it is also important to reduce the external heat leak as much as possible. For this reason the thermal shields, anchored to liquid nitrogen and liquid helium tanks, are installed surrounding the ^{3}He pot and coldfinger.

The initial version of this commercial Oxford ^{3}He cryostat was not suited for the experimental purposes. Cooling down the massive setup (about 1 kg) that is used for our experiments, only the base temperature of 0.9 K could be reached with a cooling cycle of about 2 h. These results were not satisfactory nor in temperature nor in time scales. For this reason the whole cryogenic

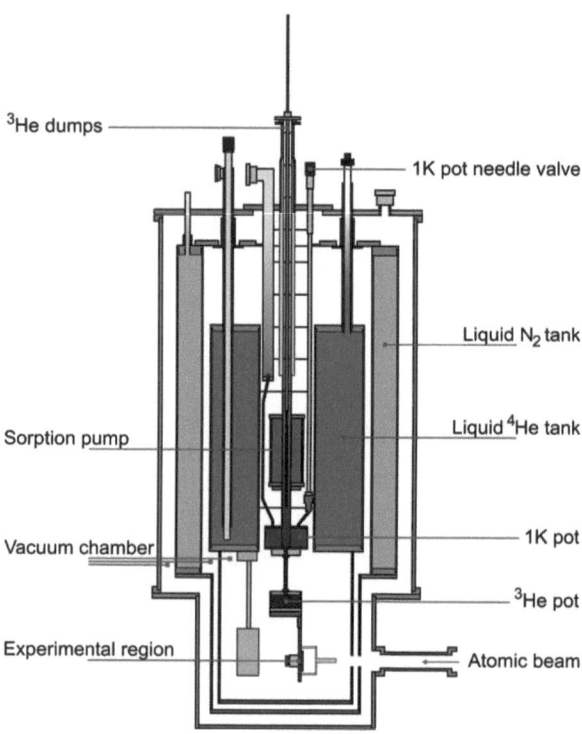

Figure 3.18: The basic principle of the ^{3}He cryostat. By means of evaporation of ^{3}He the experimental setup is cooled down to temperatures of several hundred mK.

3.6. ^{3}HE CRYOSTAT

setup was modified and afterwards the temperature of 0.3 K with a cooling cycle of more than 12 h was achieved.

The reason for this were mainly the problems related to the excessive heat conduction, radiative heating, and cooling power that was insufficient. The problems were solved addressing the various points individually:

1. Heat conduction: the thermal load from outside the cryostat through the electrical cooper wires which go to experimental region have been reduced by substituting them into the manganin ones, which have lower thermal conductivity. Furthermore, these wires are thermally anchored at the liquid N_2 and ^{4}He tanks.

2. Radiative heating: in the thermal isolation shields instead of the holes for the atomic and laser beams the small 5 mm of diameter and 30 mm long tubes have been installed reducing thermal radiation from outside, especially in the microwave regime. Furthermore, the liquid nitrogen and liquid helium temperature thermal isolation shields were covered with a thermal superinsulation foil also.

3. Cooling power: to reach cooling better performance, the partial pressure and amount of ^{3}He in the cryogenic setup was changed. Additional ^{3}He gas was added to the system, which allows now to condense about 42 cm^3 of ^{3}He in the pot. Since the sorption originally was designed to absorb much lower amounts of condensed ^{3}He (about 20 cm^3), additional mechanical pump was connected to the closed ^{3}He system. This pump is used to pump on the ^{3}He prior to the sorption pump reaching some tens of mbar pressure. In this way the coldfinger is cooled down slowly in more efficient way, letting all experimental equipment to reach thermal equilibrium.

4. Weight reduction: some components of the experimental region like atomic beam collimator, cavity squeezing mechanism and the field-ionization detection system have been re-designed to reduce the mass which has to be cooled.

After all these implemented improvements, the performance of the system has dramatically improved. In the end, the base temperature of 0.3 K reached with cooling cycle of more than 12 h.

Chapter 4

Measurements with the Atomic Beam

4.1 Magnetic Field Compensation

Magnetic field should be compensated for reduction of magnetic field flux frozen inside the superconductive cavity and for avoidance of Zeeman level splitting.

In the one-atom-maser system, magnetic field should be compensated once before cooling to superconductive temperatures due to two reasons:

- to reduce magnetic field flux frozen inside the superconductive cavity which worsens its Q-factor;

- to avoid Zeeman level splitting under which the operation of the two-level one-atom-maser is hardly possible.

The magnetic field is compensated once before every measurement cycle. Then the resonator is cooled down below the superconductive temperature and the magnetic field state is "frozen" inside. Such a situation remains during the whole cooling cycle, i.e. several weeks.

Due to the technical reasons, it is not possible to perform such a magnetic field compensation inside the resonator using usual magnetic sensor equipment, so in the micromaser the atoms themselves are used as the magnetic field probes. The main scheme for magnetic field compensation is shown in Figure 4.1. Compensating magnetic field is produced by three pairs of perpendicular Helmholtz coils. These are arranged outside the cryostat so that the resonator lies in center of the symmetry point.

With the help of circularly polarized light of the first laser (780 nm), the magnetic moment of all atoms are polarized along one direction by means of optical pumping. Afterwards the atoms move through the resonator where magnetic field should be compensated. During the time of passage the hyperfine structure spin of the atoms precess around magnetic field thereby by

52   CHAPTER 4. MEASUREMENTS WITH THE ATOMIC BEAM

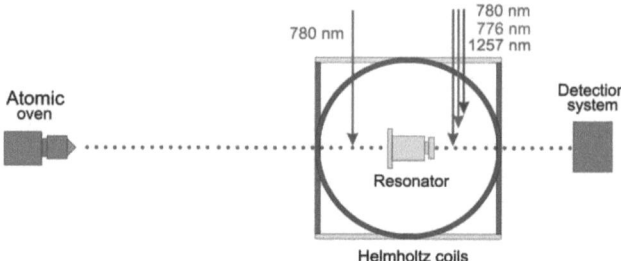

Figure 4.1: Magnetic field compensation scheme.

changing the distribution of the initially prepared quantum number state. This means that in non vanishing magnetic field, the atomic state will evolve to superposition of different m_f states. Each part of corresponding component is well defined and depends on the time of passage through the magnetic field and it's strength. In the end, using three step excitation scheme atoms are promoted to the Rydberg state, where they are detected. Due to the different Clebsch-Gordan coefficients, the excitation probability differs for each m_f transition. The experimental results of such a magnetic field compensation measurement in each of the three directions is presented in Figure 4.2. In this case, magnetic field compensation is done with the accuracy of less than 1 mG, which is several times better than it was achieved in the past.

The detailed description of this improved magnetic field compensation by using new three step excitation scheme with the thermal atomic beam and the new developed method achieving even several orders of magnitude better compensation accuracy (~ 10 µG) by using demodulation technique will be done separately in Chapter 6. There is also a new setup and experimental

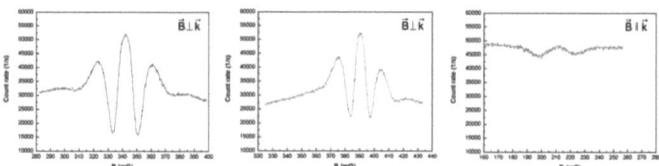

Figure 4.2: Experimental magnetic field compensation curves in all the three directions

4.2 Resonance Curve of the Maser

results of coherent atomic control by means of the Hanle precession in static magnetic field with a well-defined-velocity atoms that will be presented.

4.2 Resonance Curve of the Maser

To check the maser operation, the resonance curve (so called "maser line") of the maser is taken. The cavity is slowly tuned over the maser transition by changing the voltage of the piezoelectric transducer. Simultaneously, the field-ionization signal is recorded.

The main experimental parameters are presented in the following Table 4.1:

Table 4.1: Main experimental parameters

Maser transition:	^{85}Rb $63P_{3/2} \rightarrow 61D_{5/2}$
Transition frequency:	21.456 GHz
Lifetime $63P$ ($61D$):	488 µs (244 µs)
Coupling constant $\frac{g}{2\pi}$:	7 kHz
Cavity Q-factor:	$1 \cdot 10^9$
Cavity temperature:	0.5 K

The count rate in two channeltrons corresponding to the two states (upper $63P_{3/2}$ and lower $61D_{5/2}$ respectively) starts to change in the vicinity of the resonance frequency (21.456 GHz). Atoms in the upper maser state begin to exchange the energy with the cavity field leading to the superposition between the two states due to Rabi oscillations phenomenon (see Section 2.5). Measured maser line for the count rate of 350 counts/s ($N_{ex} \approx 10$ (number of atoms per cavity decay time)) is shown in Figure 4.3.

The observation of maser-lines has a long history in experimental micromaser physics. The first work, that demonstrated the stimulated emission with at most one atom in the cavity was based on the observation of saturation broadening of maser lines [Mes84] [MWM85]. However, a detailed theoretical analysis of the micromaser was published only later. Furthermore, the argumentation in this paper was rather qualitative. In a later work [Rai95] [RBW95], large deviations of observed maser-lines from expected curves were observed for large atomic fluxes using velocity selected atoms. These effects were explained by a complicated stray-field effect theory in the entrance and exit holes of the cavity.

In our experiments a rather simple theoretical model for the maser line width and its shape explanation was used. The experimental results of such maser line done at 1.6 K temperature for different N_{ex} together with the

54   CHAPTER 4. MEASUREMENTS WITH THE ATOMIC BEAM

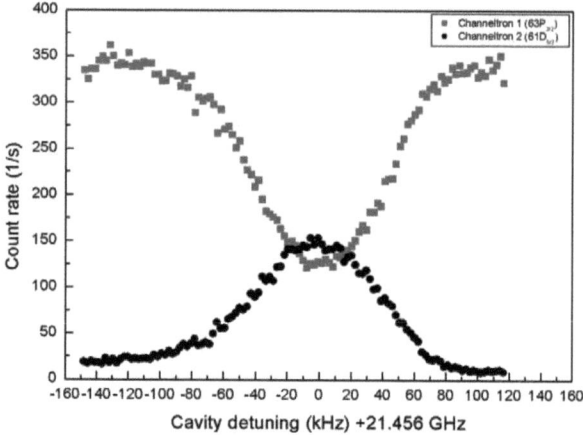

Figure 4.3: Resonance curve of the maser. The count rate in the two channeltrons changes as the frequency of the resonator is tuned over the resonance.

corresponding theoretical fit is shown in Figure 4.4. For the measurement of the maser lines a thermal atomic beam was used where all the velocities are averaged. The features of the maser line and its width for different atomic rates N_{ex} can be explained using the approach based on the steady state equation (see Section 2.4) of the micromaser. Two things were taken into account:

1. Rabi frequency term is substituted with the detuned Rabi frequency term:
$$P_g = \frac{|\Omega|^2}{\Omega'^2} \sin^2\left(\frac{1}{2}\Omega'\tau\right) \quad (4.1)$$
where $\Omega'^2 = \Omega^2 + \Delta^2$ is effective Rabi frequency.

2. Atomic velocities are averaged over many oscillation periods of the Rabi frequency:
$$\lim_{T\to\infty} \int_{-T/2}^{T/2} \sin^2\left(\frac{1}{2}\Omega'\tau\right) d\tau = \frac{1}{2} \quad (4.2)$$

This is a rather simplified approximation because it requires that a really broad velocity range is covered in curve, which is not always the case. How-

4.2. RESONANCE CURVE OF THE MASER

ever, the nice coincidence with the measured data justifies it. At higher count rates in the cavity builds up a photon field which leads to maser line broadening. A slight asymmetry in some maserlines is still observed which can be attributed to entrance hole effects, but especially on the high frequency wing of the maser line the agreement between experiment and theory is excellent.

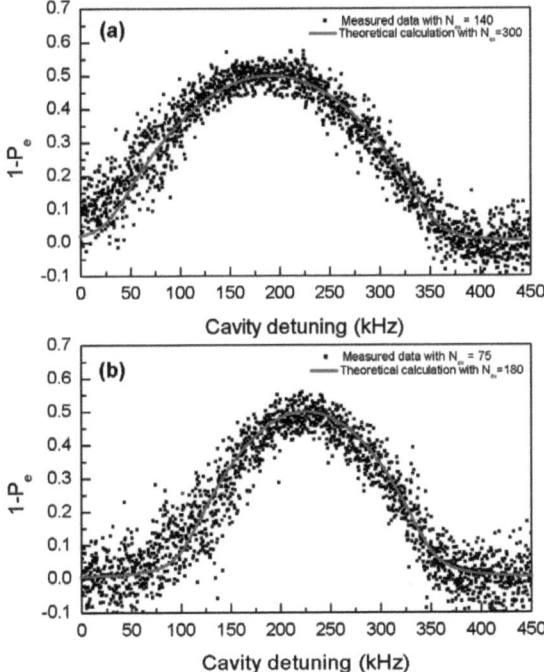

Figure 4.4: The experimental resonance curve measurements for the two different count rates: a) $N_{ex} = 140$ and b) $N_{ex} = 75$. Fitted theoretical curves corresponds to a higher N_{ex} numbers, correspondingly $N_{ex} = 300$ and $N_{ex} = 180$ from where we extract the detector efficiency of about 45 %.

By using an atom field coupling constant of $g/2\pi = 7\,\mathrm{kHz}$ which has been independently determined in a high resolution vacuum Rabi oscillation measurement (see Section 4.4), an overall detection efficiency of the Rydberg atom detection process of ∼ 45 % can be extracted by fitting the calculated theoretical curves to the experimental results (Figure 4.4 and 4.5). The atom field coupling constant is the only free parameter used to calculate the maser lines. This number gives the ratio of the probability that atoms are really detected compared to the number of atoms that pass through the resonator. This extracted detection efficiency of ∼ 45 % includes all the loss mechanisms taking into account the decay of the Rydberg states going from the excitation to the detection region, the ionization process and finally the cooled detectors (see Section 3.4) which have lower detection efficiency (unfortunately not specified by the manufacturer) than at the room temperature. Such a detection efficiency estimation from the theoretical maser line width model was done for the first time in the maser experiments.

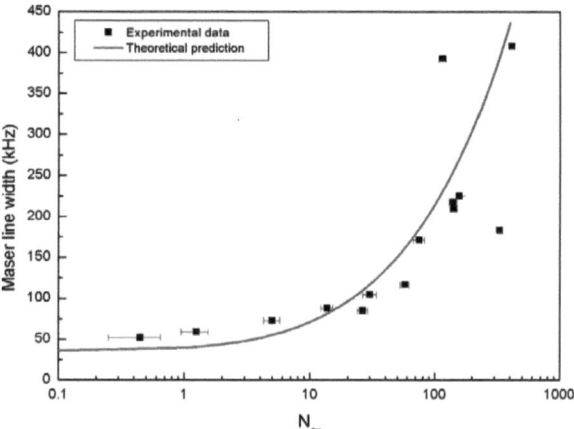

Figure 4.5: The resonance curve linewidth dependence on the count rate. To the theoretical predicted fit plotted experimental data includes previously from the measurements extracted (see Figure 4.4) ∼ 45 % detection system efficiency.

Previously in the maser experiments also some measurements were done where the detector efficiency of 19 % was extracted [Bab89], but there the comparison to incomplete theoretical model without a maser line was used and the atom-field coupling constant g was not determined in the independent measurements.

Another effect that was observed by measuring the maser lines with

4.3. VELOCITY SELECTION AND TOF MEASUREMENTS 57

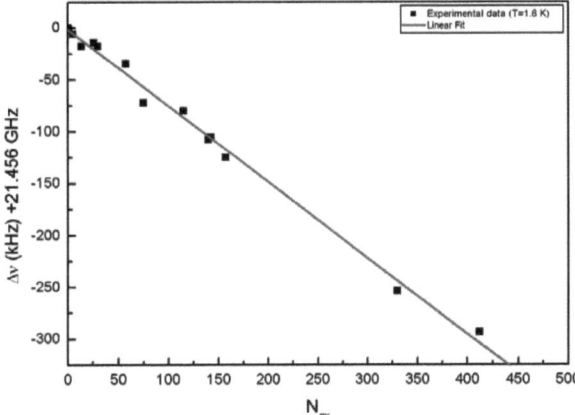

Figure 4.6: The maser line resonance shift dependence on the count rate.

different pump rates N_{ex} is the resonance shift to the lower frequencies as the number of atoms passing through the resonator is increased. The experimental results are shown in the Figure 4.6. The observed linear shift is compatible with the measurements described in [Rai95] where this effect is explained in the terms of stray field theory in the entrance holes of the resonator. The measurements confirms the predictions of this model by observing a linear shift of the resonance peak as a function of the mean photon number with a slope of 0.73 kHz per photon. Such linear character of this shift was not demonstrated in earlier experiments. Whereas this theory still remains the most complete theoretical work to explain all features on maser lines, its complicated nature hinders intuitive understanding of the physical processes in the cavity. These effects will not be discussed here, since in the experiments all the following measurements are done at very low count rates ($N_{ex} \leq 1$), where the shift is negligible and simple micromaser theory dominates.

4.3 Velocity Selection and Time of Flight Measurements

Accurate control of the interaction time between atom and field plays an important role in the one-atom maser experiments. It differs thereby strongly

58   CHAPTER 4. MEASUREMENTS WITH THE ATOMIC BEAM

from the usual maser, where the uncertainty of this parameter averages out the typical microscopic effects. The interaction time is equal to the transit flight time of the atoms in the the resonator, therefore there is a need for a good atomic velocity distribution selection.

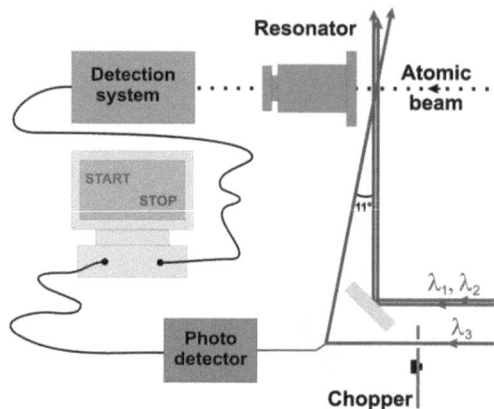

Figure 4.7: Time of flight measurement scheme.

In the experiment, the velocity of the atoms is selected by means of the Doppler effect using angular excitation. In this case only a certain velocity group of the atoms will be excited, whereas the rest would pass through the experimental region in the ground state. To avoid the contact potentials and to shield the region from the external stray fields, the excitation of atoms is done in a niobium cylinder shortly before the resonator. The first two laser stage beams hits the atomic beam perpendicularly (to match the corresponding transition frequencies since they are stabilized in the Rb cells on the Doppler free transition signals, as described in Chapter 5), whereas the third stage laser beam is superimposed on the atomic beam at the same point but at an angle of approximately 11° (in special case, for coherent atomic control measurements, the third laser beam can be superimposed along the atomic beam, i.e. at 0°). Due to the Doppler effect, the atoms see the third stage laser beam blue-shifted, with the frequency:

$$\nu' = \nu_l \frac{c + v \sin \theta}{c} \qquad (4.3)$$

where ν' is the Doppler shifted frequency, ν_l is the frequency of the third stage laser beam, v is the velocity of the atom, θ is the angle between

4.3. VELOCITY SELECTION AND TOF MEASUREMENTS

this laser and atomic beam and c is the velocity of light. By changing the frequency of the third laser (the frequency tuning is described in the Section 3.4), it is possible to select a certain group of the atoms with the corresponding velocity:

$$v = \frac{c}{\nu} \frac{\Delta\nu}{\sin\theta} \qquad (4.4)$$

where ν is the frequency of the transition $5D_{5/2} \rightarrow 63P_{3/2}$ and $\Delta\nu = \nu - \nu_l$. The velocity selection resolution depends on the atomic beam divergence, laser linewidth and the excitation angle. The three step diode system in this sense has a drawback compared to the previously used dye laser, because only from the laser wavelength (1257 nm and 297 nm correspondingly) one gets the difference in the uncertainty of the atomic velocity selection of about 4 times. In addition, the role plays also the laser linewidth, which for IR diode laser is also about 2 times larger as for UV dye laser (2 MHz and 1 MHz [1] correspondingly).

Typically the values of $\delta\tau/\tau = 8\%$ for the velocity selection are reached. The relation between the time of flight t_{tof} from the excitation region to the detection and Stark voltage U_{St} for the atoms with the velocity v can be written as

$$t_{tof} = P_1 + P2/U_{St}^2 \qquad (4.5)$$

The values P_1 and P_2 are system dependent values, which should be determined from day to day repeatedly, because they depend on not so exactly controllable parameters, like the excitation angle and stray fields at the Stark plates caused by Rb-deposition. Such a calibration of the P_1 and P_2 values is made by time of flight measurements. This is realized by using pulsed laser excitation where the time of passage of the atom from excitation region till the arrival to detection system is measured. Such measurement setup is shown in Figure 4.7.

For the generation of the uniformly time spaced laser pulses from the continuous laser beam, a mechanical beam chopper is used. Compared to the optical laser beam modulators (EOM) the mechanical chopper has the merit concerning the extinction ratio, which play an important role in maser experiments to avoid the excitation events, that spoils the measurement statistics. However, mechanical beam chopping has also a drawback wherein one cannot choose flexibly the pulse duration and separation times. For the time of flight measurements the corresponding laser pulse and separation times are chosen that the interval between two consecutive laser pulses is much larger than the average passage time of the atom from the excitation point to the detection. Moreover, it is important that the pulse width of the laser gives a reasonable compromise between the number of excitation events and resolution. Therefore, the following values were chosen: laser pulse width is 5 µs (FWHM) and the separation between the pulses is 20 ms.

---
[1] Laser linewidth data were taken from the corresponding product specification sheets.

60 CHAPTER 4. MEASUREMENTS WITH THE ATOMIC BEAM

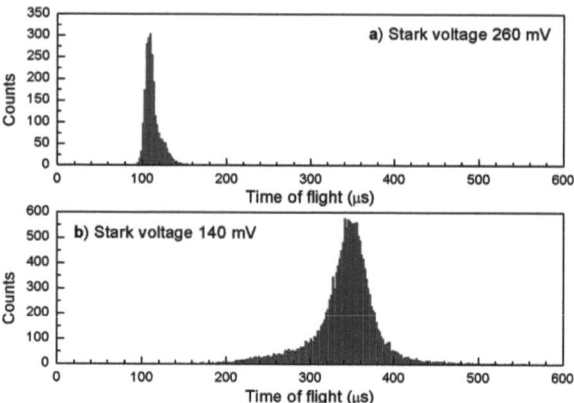

Figure 4.8: Time of flight measurement for two different Stark voltages.

As shown in Figure 4.7, mechanically chopped third laser beam at an angle of 11° crosses the other two laser stages and atomic beam just before the resonator (detuned). A part of the splitted third laser beam goes to the photodiode which gives a reference signal for the start of the measurement when the atom is excited. The time of flight of the atom is measured until it reaches the detection system, where corresponding click gives the signal to stop the measurement. For a calibration, several such a time of flight spectra are taken and for each time interval occurrence histogram is made. Figure 4.8 shows two of such a measurements for different Stark voltages.

The distribution width of the arrival times in the time of flight spectrum can be explained by the Doppler equation which shows that the distribution of arrival times is expected to be broader for the slower atoms:

$$\delta t = \frac{x}{v^2} \frac{c}{\sin \theta} \frac{\delta \nu}{\nu} \qquad (4.6)$$

The different side slopes can be explained in the following way: the laser linewidth is symmetric about some chosen value in the thermal atomic velocity distribution profile. Now depending on which side of this thermal velocity distribution profile one selects the atomic velocity group, on one side there will be more atoms than in the other one. If one doesn't saturate and assume some constant excitation probability, then the excitation probability would be proportional to the number of atoms with exact that velocity,

4.4. VACUUM RABI OSCILLATIONS

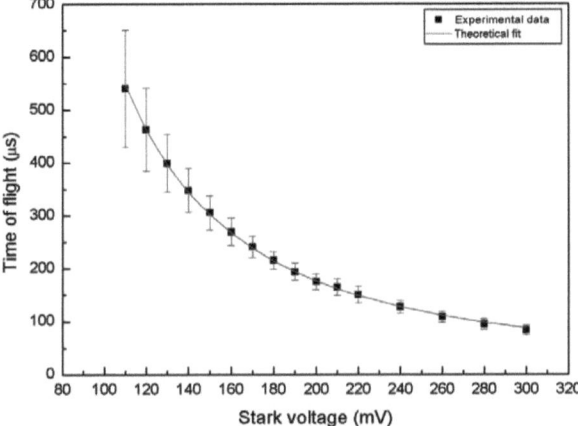

Figure 4.9: Time of flight calibration measurement. The width of the error bars is given by the widths (FWHM) of the velocity distributions, as shown in Figure 4.8.

which correspond to the laser detuning. The product of the laser linewidth and the thermal velocity distribution profile will get different slopes of selected atomic velocity distribution profiles, as shown in Figure 4.8.

For the velocity distribution calibration the time of flight spectra for different Stark voltages are measured. Experimental results are shown in Figure 4.9. The theoretical fitted parabola is using the equation (4.5) to these time of flight spectra adapted curve.

4.4 Vacuum Rabi Oscillations

As discussed in the Chapter 2, the interaction of a two-level atom with a single mode of the cavity field is described by the Jaynes-Cummings hamiltonian. In this model, an atom in the presence of a resonant quantum field undergoes Rabi oscillations. If the cavity contains the quantum mechanical vacuum field, a single atom that enters the cavity in state $|e\rangle$ will leave it in the same state with probability

$$P_e = \cos^2(|g|\tau) \quad (4.7)$$

where g is atom-field coupling constant and τ is interaction time.

62   CHAPTER 4. MEASUREMENTS WITH THE ATOMIC BEAM

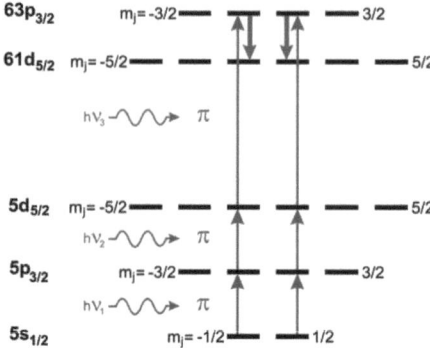

Figure 4.10: Measurement scheme of vacuum Rabi oscillation.

At zero atomic flux, the cavity contains the blackbody field. The average number of thermal photons at 0.65 K temperature is $\bar{n}_{th} = 0.26$. When an atom in the upper maser level enters the resonant cavity, the probability of spontaneous decay is increased as a result of the enhanced vacuum field in the cavity. In addition, the emission is stimulated by the thermal radiation field, which is, as mentioned above, quite small. As a result of this emission, the cavity field is increased by the emitted photon, and the atom, now in the lower maser level can reabsorb a photon and return to the upper maser level. Since the average number of photons accumulated by the Rydberg atoms in the cavity is approximately given by $\bar{n}_m \approx \tau_f r/2$, where r is the number of Rydberg atoms passing through the cavity per one second, for vacuum Rabi oscillation measurements r was chosen to be $\sim 3\,\mathrm{s}^{-1}$, to ensure that the cavity is in the vacuum state. With $\tau_f = 73\,\mathrm{ms}$ (see Section 3.5) this gives a $\bar{n}_m \approx 0.05$, what is on the secure side for observation of pure vacuum Rabi oscillations, even when the probability of detecting an atom in either the excited or ground state is about 45%, as it was mentioned in the Section 4.2. Due to the finite lifetime of the atoms, additional atoms are lost through spontaneous decay in the flight between the cavity and the detectors.

The experimental setup for measurements of Rabi oscillations is equivalent to one of the time of flight measurement, which was depicted in Figure 4.7. Pulsed excitation of the atoms is used so that the number of atoms passing through the cavity can be predetermined. The polarization of the laser beams were chosen accordingly as depicted in Figure 4.10, where only the $63P_{3/2}$ $m_j = \pm 1/2$ levels are populated. This is a conventional setup

4.4. VACUUM RABI OSCILLATIONS

which has been always used before in maser experiments with one step laser excitation, as the same scheme is realized also with the three steps.

When the field-ionization signal is recorded in the experiment, the contribution of many atoms is averaged. Changing the selected velocity leads to a different interaction time and leaves the atom in another phase of the Rabi cycle when it reaches the detector. Measured atomic inversion

$$1 - 2P_e = -\cos(2|g|\tau) \qquad (4.8)$$

(here P_e is the probability to detect an atom in the upper maser state) oscillates sinusoidally around zero and is called vacuum Rabi oscillation. The experimental results are shown in Figure 4.11. Measurement using

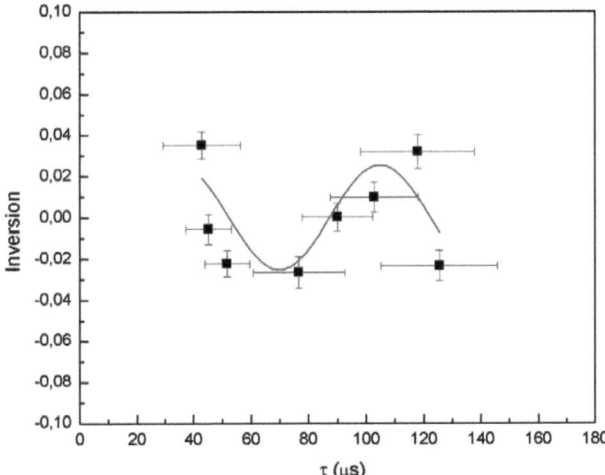

Figure 4.11: Experimental results of vacuum Rabi oscillation measurement. From the measurement extracted coupling constant $g/2\pi = 6.7$ kHz.

such a excitation scheme is a little bit inconsistent because it leads to a statistical mixture of the $m_j = +1/2$ and $m_j = -1/2$ states that interacts with the resonator mode. Only due to the fact that the interaction for both of these two pairs is exactly the same, allows to interpret the result as the investigation of one of these two transitions.

The resolution of measured Rabi oscillations is comparable with the results achieved by Varcoe et al. [VBWW00]. Between the factors that reduces

64   CHAPTER 4. MEASUREMENTS WITH THE ATOMIC BEAM

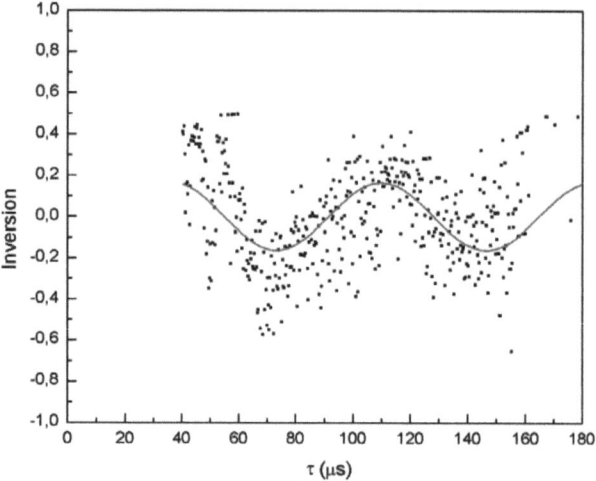

Figure 4.12: Experimental results of vacuum Rabi oscillation measurement, analyzed in a single time units, given by acquisition program.

the resolution like atom - atom collisions, detection efficiency, significant role also plays limited atomic velocity distribution selection (discussed in section 4.3) in recent experiments with the three step excitation.

From the measurement extracted coupling constant $g/2\pi = 6.7$ kHz is in a good agreement with theoretically predicted values and the previously experimentally measured values of ~ 7 kHz [Mes84] [Rem86] [Ben95] [Rai95] [Bra01]. However, most of these previously measured g-values were extracted from the measurements like maser resonance lines or pump curves, where the measurements are influenced by the detector efficiency. And the only one direct measurement which does not dependent on the efficiency of the detectors for the coupling constant g-value establishment is the vacuum Rabi oscillation measurement.

To reduce the atomic velocity selection distribution effect this measurement was analyzed in a different way also. The inversion for not whole velocity group as measured was calculated, but the whole measurement data were divided into a time scale of the resolution of the measurement program of 1 μs (even when the excitation pulse width was larger: 5 μs). Since the measurements are taken with the 1 μs scale resolution, the whole

4.4. VACUUM RABI OSCILLATIONS

measured experimental data of velocity distributions can be correspondingly overlapped and summed. In such a way analyzing the data and calculating the inversion, it is possible to avoid this velocity distribution spreading effect, and afterwards to extract the result of the Rabi oscillations for each 1 μs time interval. The final result is shown in Figure 4.12. The extracted from the measurement coupling constant, as expected, remains the same: $g/2\pi \approx 7$ kHz. However, as can be seen from the extracted data, in this case the resolution of the vacuum Rabi oscillation measurement is significantly increased up to 0.2. This result shows about ten times better contrast of such a measurement than it was achieved by predecessors ([VBWW00]).

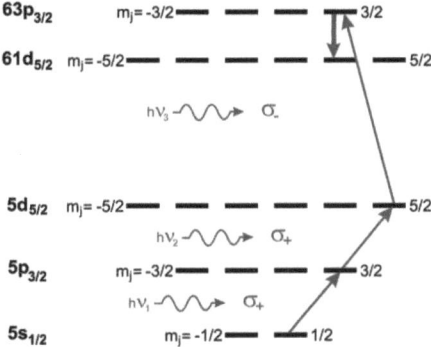

Figure 4.13: New measurement scheme of Rabi oscillation using a single transition.

Using flexibility, that three step excitation setup gives, a different excitation scheme was realized also (Figure 4.13), which allows to investigate the single $63P_{3/2}\ m_j = 3/2 \rightarrow 61D_{5/2}\ m_j = 3/2$ transition. As only one transition is involved, such a measurement is insensitive to any possible atomic levels shifts, i.e. due to the remnant magnetic fields, which was not the case in the previously described vacuum Rabi oscillation measurement. Using this single transition, Rabi oscillation measurements were done for the first time in the micromaser experiments. The experimental results of such a measurement is shown in Figure 4.14. From the measurement extracted coupling constant $g = 5.3$ kHz.

If one looks carefully at what contributes to Rabi frequency, then it is product of various quantities, and only factor that differs for these are Clebsch-Gordon coefficients for these transitions. Sice the Clebsch-Gordon coefficients can be evaluated directly, one can calculate the theoretical ratio

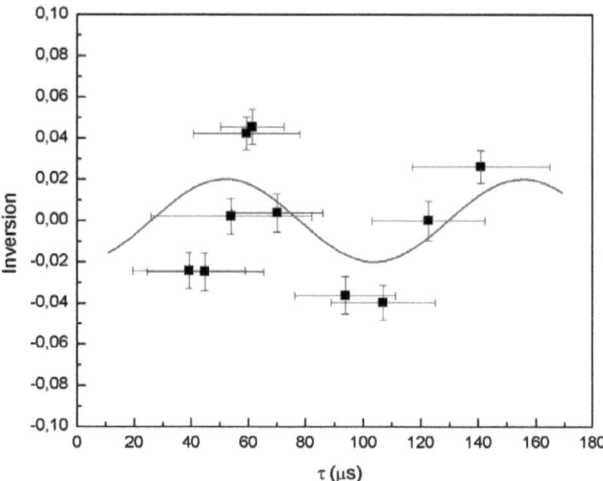

Figure 4.14: Experimental results of vacuum Rabi oscillation measurement using single transition.

using Wigner 3-j symbols for the two $m_j = \pm 1/2$ and $m_j = 3/2$ transitions correspondingly:

$$\begin{pmatrix} \frac{3}{2} & 1 & \frac{5}{2} \\ \frac{1}{2} & 0 & -\frac{1}{2} \end{pmatrix} = -0.3163 \qquad (4.9)$$

$$\begin{pmatrix} \frac{3}{2} & 1 & \frac{5}{2} \\ \frac{3}{2} & 0 & -\frac{3}{2} \end{pmatrix} = 0.2582 \qquad (4.10)$$

so theoretically predicted ratio

$$\frac{|g_{3/2}|}{|g_{1/2}|} \approx 0.82 \qquad (4.11)$$

The comparison with the experimental g-factor ratio $\frac{g_{3/2}}{g_{1/2}} = 0.79$ yields a good agreement with the theoretical prediction. This good agreement confirms, that in the previous measurement of Rabi oscillations with the $m_j = +1/2$ and $m_j = -1/2$ states (see Figure 4.10) no state degeneracy was present what would lead to a statistical mixture of these transitions and therefore different value of atom-field coupling constant g.

4.4. VACUUM RABI OSCILLATIONS

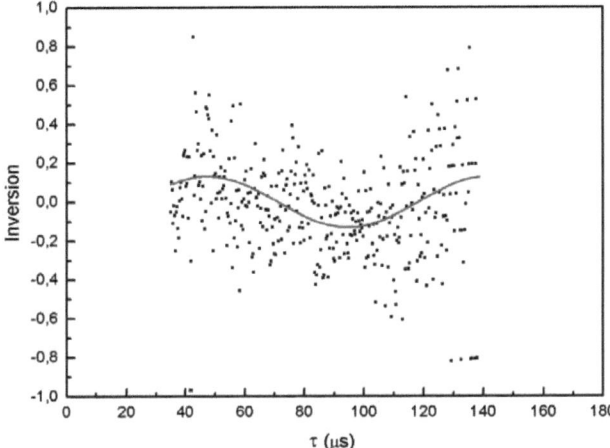

Figure 4.15: Experimental results of vacuum Rabi oscillation measurement using single transition and analyzed in a single time units, given by acquisition program.

Such measurements also lead to a conclusion that the magnetic field in the system is compensated well enough, that it doesn't influence the measurements due to Zeeman effect level splitting.

This measurement for reducing the atomic velocity selection distribution effect was also analyzed in the same way described earlier. The results are shown in the Figure 4.15. Extracted g-factor is almost the same: $g \approx 5.2$ kHz, but the resolution is increased up to ~ 0.2.

The measurement of the increased vacuum Rabi oscillation and for the first time in the micromaser experiments investigated and measured pure magnetic level transition Rabi oscillation, furthermore, the reproducible results of such a measurement shows, that with the constructed micromaser apparatus is possible to investigate the quantum effects in the resonator with almost 10 times increased resolution as it was achieved in previous micromaser experiments [VBWW00] [Bra01].

The remaining inconsistencies with the theoretical prediction of the vacuum Rabi oscillation resolution (see Figure 2.7, the curve of $T = 0.7$ K for comparison) should be attributed to still existent limited atomic velocity distribution selection and atom - atom collisions. Additional point could be higher field temperature, because with the temperature sensor one measures only the temperature of the cavity walls. In fact the field inside the resonator can be hotter, because resonator has the holes, where the outside radiation from hotter surface can come in, like from an atomic oven. Therefore could be possible, that the field temperature inside is higher.

Chapter 5

Laser System and Spectroscopy

5.1 Overview

For the efficient excitation of ^{85}Rb atoms to the upper maser state ($63P_{3/2}$), a new three step diode laser setup is used. As already mentioned in the Section 3.3, this new system was developed because the initially used frequency doubled dye laser system showed several problems, like power and beam spatial instabilities, bad laser mode quality, frequency doubler crystal degradation (deuterated ammonium-dihydrogen-arsenate crystals are no more produced) and finally high maintenance costs (everyday intracavity frequency doubled ring dye laser adjustment, weekly dye change).

In the experiments used new excitation scheme is depicted in Figure 5.1. ^{85}Rb atoms from the $5S_{1/2}, F=3$ hyperfine structure state are efficiently pumped to the Rydberg $63P_{3/2}$ state by means of three resonant steps:

$$5S_{1/2}, F=3 \longrightarrow 5P_{3/2}, F=4 \longrightarrow 5D_{5/2}, F=5 \longrightarrow 63P_{3/2}$$

The three lasers whose frequencies are stabilized on the three mentioned transitions ($\lambda_1 = 780.243$ nm, $\lambda_2 = 775.978$ nm and $\lambda_3 = 1256.730$ nm) are one with another spatially overlapped in order to accomplish this excitation scheme. The excitation probability in three steps to $63P_{3/2}$ state is also higher than in using one step excitation, since the overlap of the wave functions of the involved intermediate states in each case is much larger. Previously, the low count rates prolonged the measurements and in consequence, it made them more susceptible to long term drifts. With the efficient promotion of atoms to Rydberg states by means of three-step excitation high flux experiments also with the slow and fast atoms in the range of velocity distribution wings can be done.

Another problem with former UV laser setup was that using one step excitation in the experiments for interaction times below 40 µs due to the

70 CHAPTER 5. LASER SYSTEM AND SPECTROSCOPY

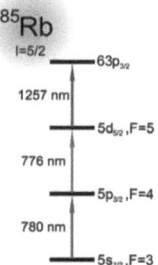

Figure 5.1: Excitation of ^{85}Rb atoms to the upper maser state scheme using three-step diode laser system.

Doppler effect the excitation spectra of $63P_{3/2}$ and $63P_{1/2}$ fine structure levels overlap, leading to excitation of $63P_{1/2}$ state atoms also, which do not interact with the resonator field however are counted in the detectors and lead to perturbation of the counting statistics. This problem is solved by using three-step laser excitation setup, where flexibility of choosing atomic selection rules is used and meanwhile allows to perform the experiments that were not possible with former one step excitation setup.

The main scheme of the three-step diode laser system is shown in Figure 5.2. For all the three stages, a commercially available external-cavity Littrow configuration diode lasers (TOPTICA Photonics AG, type SYS DL 100) are used together with voltage supply and monitor unit (type DC 100), as well as a temperature stabilization module (type DTC 100), and current control model (type DCC 100). The first two lasers (stage one: $\lambda_1 = 780,243$ nm, $P = 80$ mW and stage two: $\lambda_2 = 775,978$ nm, $P = 40$ mW) are frequency stabilized on spectroscopic signals generated in the Rb cells and the third laser (stage three: $\lambda_3 = 1256,730$ nm, $P = 25$ mW) is locked on the atomic beam signal.

The adjustment to the corresponding resonant wavelength in each of the lasers is done by a simple angular rotation of the grating relative to the optical axis of the laser cavity. With such angle tuning of the diffraction grating, a small tuning range in the order of 10 GHz may be performed. Fine tuning of the emission wavelength of such a laser diodes is performed by changing the temperature of the laser case or by changing the injection current. Changing the temperature of the laser mount is suitable for a larger wavelength tuning. Since the change of the temperature is a slow process, tuning rates are very limited. Changing the laser current results in a faster tuning, however the total tuning range is much smaller than the tuning

5.1. OVERVIEW

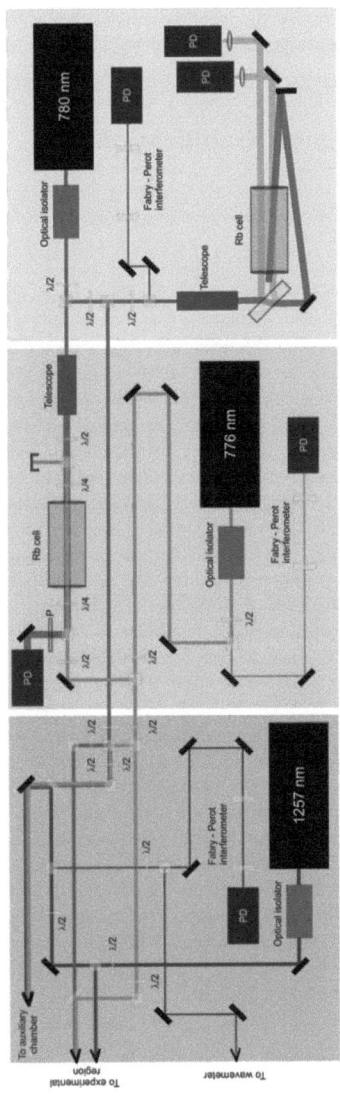

Figure 5.2: Diode laser system. It consists of the three commercially available diode lasers stabilized on the signals from the Rb cells and atomic beam correspondingly.

range by changing the laser temperature.

All the details about each of the laser stages, setup and frequency stabilization on the corresponding transitions will be discussed in the following sections.

5.2 Laser Frequency Stabilization

For the laser stabilization on the required frequency, lock-in detection scheme is used to generate the error signal which is afterwards processed with specially made PID regulator.

The laser frequency stabilization scheme used in former experiments with the dye-laser was not satisfactory because the frequency was stabilized on the fixed count rate, where the change in the atomic flux led to undesired frequency change. To overcome this problem, a new top-of-fringe laser frequency stabilization system for the new diode laser setup was developed using specially designed error detection and feedback schemes.

All the three laser stages are frequency stabilized on the spectroscopic signals from the corresponding transition lines using the same technology, namely synchronous demodulation scheme. The drifts or the instabilities in laser frequency can be minimized to zero through a feedback mechanism that monitors the output frequency of the laser and uses it to produce an error signal, with rapid enough action to force the laser frequency to remain at a fixed value. As a reference frequency for this feedback mechanism, the required hyperfine transition of Rb is used. Although there are various mechanisms that can provide suitable feedback to lock the laser, in this experiment employed mechanism incorporates a lock-in amplifier (Stanford Research Systems model SR510 for the first stage and model SR530 for the second stage).

The main scheme of such a laser stabilization circuit is shown in Figure 5.3. It consists of two main parts, lock-in-amplifier and regulator. Essentially, the lock-in-amplifier takes the derivative of the corresponding peak of spectroscopy signal and produce a dispersive signal (error signal) which is then processed by a special home made regulator with a modified PID topology. This error signal has the 0 value at the maximum of the atomic transition line, which is taken as a frequency reference, and a positive or negative value corresponding on which direction the frequency is shifted off. Depending on error signal, the PID regulator delivers correction voltage to the monitoring unit of the laser system, forcing the laser output back on the summit frequency, once it randomly drifts off it, so that the error signal is regulated to the 0 value.

The PID controller involves three separate parameters: the Proportional, the Integral and Derivative values. The Proportional value determines the reaction to the current error, the Integral value determines the reaction

5.2. LASER FREQUENCY STABILIZATION

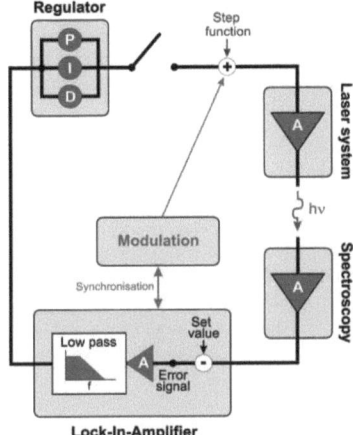

Figure 5.3: Laser frequency stabilization scheme.

based on the sum of recent errors, and the Derivative value determines the reaction based on the rate at which the error has been changing. The weighted sum of these three actions is used to adjust the process via a control element such as the laser frequency. By "tuning" the three constants in the PID controller algorithm, the controller can provide control action designed for specific process requirements. The response of the controller can be described in terms of the responsiveness of the controller to an error, the degree to which the controller overshoots the set-point and the degree of system oscillation.

The proportional, integral, and derivative terms are summed to calculate the output of the PID controller. Defining $u(t)$ as the controller output, the final form of the PID algorithm is [TQGC99]:

$$u(t) = MV(t) = K_p e(t) + K_i \int_0^t e(\tau)d\tau + K_d \frac{de}{dt} \qquad (5.1)$$

and the tuning parameters are:

1. K_p: Proportional gain - larger K_p typically means faster response since the larger the error, the larger the Proportional term compensation. An excessively large proportional gain will lead to process instability and oscillation.

2. K_i: Integral gain - larger K_i implies steady state errors are eliminated quicker. The trade-off is larger overshoot: any negative error integrated during transient response must be integrated away by positive error before reaching steady state.

3. K_d: Derivative gain - larger K_d decreases overshoot, but slows down transient response and may lead to instability due to signal noise amplification in the differentiation of the error.

The PID regulators, used in our experiment, were specially developed by G. Stania, and together with the stabilization circuit thoroughly described in [Sta05]. As an adaptation to the experimental setup some minor changes were done. Based on a mathematical model for a PID regulation scheme, the amplifications in the P, I and D sections has been determined to guarantee the operation of the system under aperiodic damping conditions. In this way, each regulator is specially adapted to the system that it controls.

Closed feedback loops are characterized by so called perturbation response. Perturbation response describes the influence of at the certain place in the closed loop (Figure 5.3, point "Step function") introduced variable disturbance on the output of the system. To test this, the stabilization circuit is closed and the step function perturbation is fed to the system. In this study, the goal of such a closed loop dynamics is aperiodic damping of perturbations. Such investigation of the system for the first and the second laser stages are shown in Figure 5.4.

The first stage implements field effect current control of the diode laser, leading to the damping times of aperiodic perturbation of about 200 μs (Figure 5.4(a)). In the second stage using the current modulation scheme, the measured perturbation damping times reaches 140 μs (figure 5.4(b)). In both cases, the aperiodic damping time constant τ is 5 times smaller than the time constant of lock-in-amplifier.

In the last, third stage, the frequency stabilization is done on the atomic beam signal, or more precisely, on the signal from the field-ionization detection system (the setup and production of signal will be discussed more in detail in Section 5.5). This detection system gives out the discrete signal, count rate, which is dependent on the laser frequency. The count rate fluctuations, which are pure statistical fluctuations, have relatively long time scales and bring noise in the system. So the regulation dynamics is limited here. Top of fringe stabilization is done using digital lock-in processing of the count rate signal and control with PI regulator. The test of stabilization for the third stage is also discussed in Section 5.5.

5.2. LASER FREQUENCY STABILIZATION

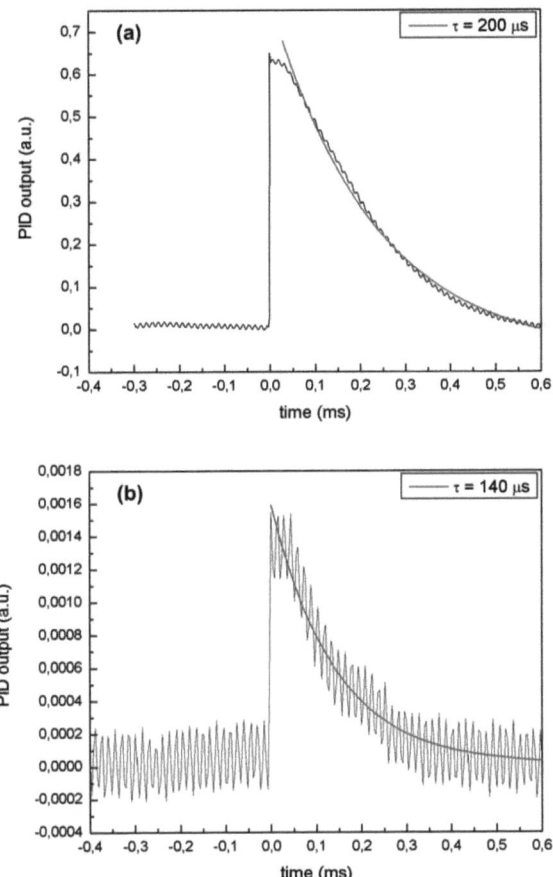

Figure 5.4: Step function response of the closed stabilization loop for the fist (a) and second (b) laser stages.

5.3 First Stage

5.3.1 The D2 Transition of Rb

As an alkali metal with one valence electron, Rb has an energy-level scheme that resembles hydrogen. The ground state electron configuration of Rb is $[1s^2, 2s^2, 2p^6, 3s^2, 3p^6, 3d^{10}, 4s^2, 4p^6], 5s^1$. As for the core Rb^+ ion, it is spherically symmetric, resulting to a total angular momentum (L), spin angular momentum (S), and thus spin-orbit $L-S$ coupled angular momentum ($J = |J| = |L + S|$) of zero. Consequently, with all the core ion quantum numbers equal to zero, the observed energy transitions with the associated changes in $L-S$ coupled quantum numbers comes from the valence electron only. As stated in $L-S$ coupling notation: $(nl)2S + 1Lj$, (where n and l are the principle and angular momentum quantum numbers respectively), the ground electronic state is $5^2S_{1/2}$, and the first electronic excited state is found in the 5^2p orbital. Here $S = 1/2$ and $L = p = l = 1$, leaving two possible values for J (i.e., $|L-S|, \ldots, J, \ldots, |L + S|$ in integer steps) equal to 1/2 and 3/2. As a result, two possible energy levels exist for these given n and l values. The first, less energetic excited state is referred to as the $D1$ line and has the following quantum numbers: $5^2P_{1/2}$. The second, more energetic excited state is referred to as the $D2$ line and has these quantum numbers: $5^2P_{3/2}$. This splitting of the $S = 1/2$ and $L = 1$ state into two finer ($J = 1/2$, and 3/2) states is known as the magnetic fine structure states of the atom, where the former state corresponds to a transition from the ground state with a wavelength of 794.7 nm, whereas the latter transition corresponds to a wavelength of 780 nm. We concentrate on the $D2$ line.

The spin quantum number of the nucleus (I) and the nuclear quadrupole moment lead to even finer splittings in the energy spacing of the atom, known as the atomic hyperfine structure. In zero or even very weak magnetic fields, I and J couple together and lead to what is known as the grand total angular momentum quantum number of the whole atom: $F = |F| = |I + J|$, where $|I-J|, \ldots, F, \ldots, |I + J|$ in integer steps. Now, it must be noted that ^{85}Rb has $I = 5/2$, whereas ^{87}Rb has $I = 3/2$. Therefore, considering only the $D2$ line where $J = 3/2$, the two isotopes each have differing states. That is, with $J = 3/2$ and $I = 5/2$, ^{85}Rb has the following possibilities for F states: $F = 1, 2, 3$, and 4, whereas with $I = 3/2$, ^{87}Rb has these following values of F states: $F = 0, 1, 2$, and 3. The ground state quantum numbers for ^{85}Rb are $J = 1/2$, and $I = 5/2$, resulting in the possible F states: 2 and 3, whereas the ground state quantum numbers for ^{87}Rb are $J = 1/2$ and $I = 3/2$ resulting in these possible F states: $F = 1$ and 2. Evidently, in both cases the ground state is split into two hyperfine levels, making the total number of possible transitions in each isotope, from the ground state to the excited $D2$ line, not eight, but six. This is due to the fact that transitions

5.3. FIRST STAGE

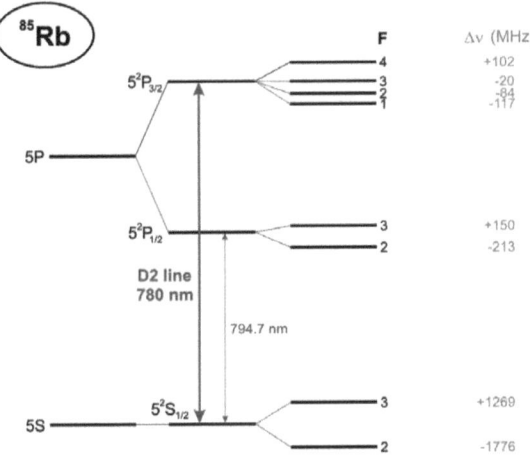

Figure 5.5: Energy manifolds of ^{85}Rb transitions.

from one energy level to another must obey certain quantum mechanical selection rules that prohibit two of the transitions above. These selection rules for the exchange of one value of quantum number for another in a given transition, for the quantum numbers F, J, L and M, dictate that no quantum number may change by more than 1. In other words, the difference between the value of a ground state and an excited state quantum number can either be 0 or 1. The only exception is that L can only change by one, not by zero. Thus, in the case of transitions from the F values of 2 and 3 for ^{85}Rb, and 1 and 2 for ^{87}Rb, to the four corresponding excited state F values of each isotope, there can only be six possible transitions as illustrated in the hyperfine energy manifold of ^{85}Rb in Figure 5.5.

5.3.2 Doppler-free Saturation Spectroscopy

The first stage diode laser is frequency stabilized on optical doppler-free saturation spectroscopy signal from Rb cell.

Doppler broadening is usually the dominant contribution to the observed width of lines in atomic spectra, at room temperature. The techniques of Doppler-free laser spectroscopy overcome this limitation to give much higher resolution.

If the radiation from the laser, oriented along the z axis, has a frequency

ν_L in the lab frame, atoms moving at velocity v_z with respect to the lab frame will see a frequency ν_A such that

$$\nu_L = \nu_A \left(1 + \frac{v_z}{c}\right) \tag{5.2}$$

When ν_A equals one of the resonant frequencies of the atom, ν_0, the atom will absorb the radiation to jump to an excited state. This means that atoms moving towards the laser, will absorb at a frequency that, when measured in the lab frame, is lower than their resonant frequency, and atoms moving away from the laser will absorb at higher than their resonant frequency. The measured absorption spectrum will thus be characterized by broad signals around each absorbed frequency. If the atom velocities are given by a Maxwell distribution, the full-width at half maximum amplitude (FWHM) of the Doppler-broadened signal is given by

$$\nu_{FWHM} = 2\frac{\nu_0}{c}\left(\frac{2kT}{M}\ln 2\right) \tag{5.3}$$

which is about 500 MHz for rubidium.

The primary feature of saturation laser spectroscopy is its avoidance of the Doppler effect. The saturation spectroscopy setup is depicted in Figure 5.6. In this technique three laser beams are utilized, all split off from the same laser and thus have the same frequency. All three beams are directed into Rb vapor cell (about 10 cm length). The Rb cell was kept at room temperature and enclosed by a μ-metal magnetic shield by which the residual magnetic field was reduced to about 1 mG. This shield is essential for obtaining large and stable signals, because a stay magnetic field is disruptive in the experiment [YUM+03]. Concerning the laser beams, there is one saturating, intense laser beam called the pump beam and two other probe beams. One beam is overlapped with the pump beam such that the two beams propagate in opposite directions and overlap at the possible smallest angle as they pass through the cell. The purpose of the saturating beam is to saturate the transition of those atoms in its path such that the overlapping probe beam cannot interact with them. The reference probe, meanwhile is displaced far enough from the pump beam within the vapor cell such that the saturating beam cannot impinge upon it. This is for the purpose of providing a reference frequency sweep. Afterwards, the two probe beams are detected by the photodiodes (PD) in producing the Doppler-free spectrum. The Fabri-Perot interferometer was used to diagnose the spectral properties of the laser and mode hop-free-scan range optimization at the required wavelength.

Now, once the counter-propagating pump beam overlaps one of the probe beams, the situation remains relatively unchanged with respect to the absorption profile, since the two beams interact with different velocity classes

5.3. FIRST STAGE

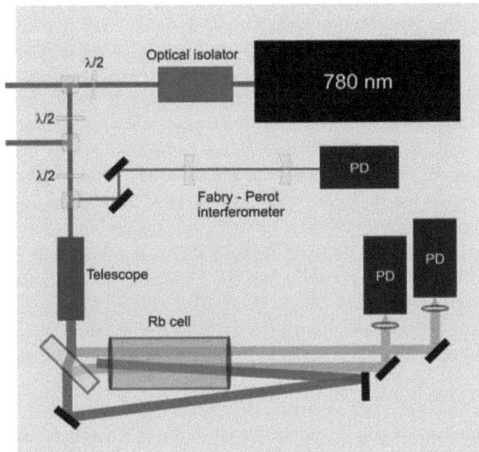

Figure 5.6: Setup of Doppler-free saturation spectroscopy.

of atoms. That is, since the two beams are of the same frequency, they each interact with atoms moving with similar speeds and thus similar Doppler transition frequencies. However, since the beams are counter-propagating, they interact with atoms with opposite directions of movement. At one particular value of frequency in the sweep, each of the two beams interacts with atoms on opposite sides of the Doppler profile. The absorbed light in both cases is either blue-shifted or red-shifted. Assuming it is blue-shifted, once the sweep goes up in frequency, the two beams switch sides of the Doppler profile, and the respective atoms now see the light red-shifted. This is just another way of saying that the production of the Doppler peak is independent of the direction of the laser beam in the lab. Since the two beams interact with a different set of atoms altogether, they have no effect on one another. As a result, the probe beam reports a signal to the PD that is unperturbed by the presence of the pump beam.

However, when the laser is nearly tuned to the lab frame or zero velocity transition frequency during the scanning, the two beams have no choice but to interact with the same set of zero velocity atoms. This results to the reduced absorption of the probe beam at one spot along the Doppler broadened profile. Namely, during the course of a sweep across an absorption peak, the PD reports the familiar Doppler broadened profile, but the absorption is suddenly reduced and the intensity goes up at one location once the sweep reaches the transition frequency of the zero velocity atoms. Once the laser is tuned to the lab frame transition frequency, both beams interact with the same atoms, but the pump beam, being much more intense, has a much larger probability of interaction. Once an atom is excited, it cannot be re-excited until it decays. In other words, once the pump beam excites a certain atom, it is removed from the pool of excitable atoms available to the probe beam. Furthermore, once the two beams are overlapped, the absorption of the probe beam is reduced by a factor governed by the number of zero velocity atoms that are excited by the pump beam. As a result, the signal reported by the overlapping probe beam tuned to the lab frame transition reveals Doppler broadened peaks with small dips (called Lamb dips)[Dem92] corresponding to the absorption frequencies of the zero velocity class of atoms only. Since the un-broadened absorption profile is a direct result of the natural linewidth of the atomic transitions, which have a Lorentzian lineshape, these Lamp dips inherit the very same Lorentzian lineshape.

Figure 5.7 shows such a saturation spectroscopy signal. A green line shows Doppler broadened signal, where the pump beam was not present and the other curve (blue) shows usual Doppler-free saturation spectroscopy signal where the all three $(5S_{1/2}, F = 3 \longrightarrow 5P_{3/2}, F = 2, 3, 4)$ hyperfine transitions are nicely resolved. Since the signal with the presence of the pump beam still has the Doppler broadened background, this can be removed by electronically subtracting the Doppler broadened background from the saturated signal. The result is depicted in Figure 5.7(b). The linewidth is given by $\delta\nu = \Gamma/2\pi$, which is the "natural width" of the transition. In the present case of Rb, $\delta\nu = 5, 98$ MHz [MvdS99].

Pressure broadening can also be a limiting factor due to the fact that at a high enough vapor pressure within the vapor cell, perturbations in Rb energy levels are manifested through the increased collisions in moving atoms. However, since the commercial Rb cell at room temperature is used, the pressure in the Rb vapor is just saturated vapor pressure, high enough for adequate signal production and low enough to avoid pressure broadening.

The final type of broadening is power broadening. In this case the laser light saturates the transitions of the atoms. A good balance of pump and probe beam power for observation of minimized linewidths is described by [PBH+80]. Here is shown that differential lineshape $A_{12}(\nu)$ for saturated absorption by a low-density target is

$$A_{12}(\nu) = A_s \left(\frac{\Delta}{2}\right)^2 \left[(\nu - \nu_0)^2 + \left(\frac{\Delta}{2}\right)^2\right]^{-1} \quad (5.4)$$

where A_s scales as

$$I_{probe} \frac{I_{pump}}{I_{sat}} \frac{1 - \Gamma_i T}{2 + \Gamma_i T}[F + F^2] \quad (5.5)$$

5.3. FIRST STAGE

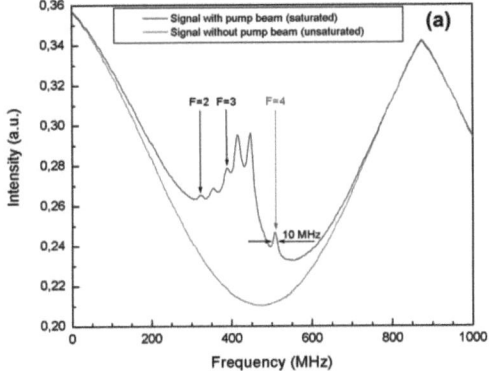

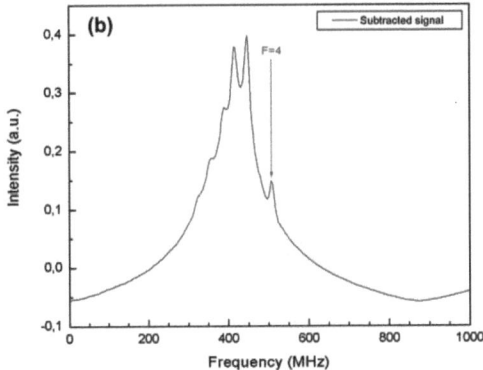

Figure 5.7: Hyperfine transitions for the first stage. a) Green curve shows Doppler broadened signal, where the pump beam was not present and the blue curve shows usual Doppler-free saturation spectroscopy signal where the all three $5S_{1/2}, F = 3 \rightarrow 5P_{3/2}, F = 2, 3, 4$ hyperfine transitions are nicely resolved. b) To get rid of Doppler broadened background, this is removed by electronically subtracting it from the saturated signal.

and $\Delta = \frac{\gamma(1+F)}{4\pi}$ is the power broadened width (FWHM) of the signal in Hz. The quantities I_{probe} and I_{pump} are laser beam intensities, $F = \sqrt{1 + \frac{I_{pump}}{I_{sat}}}$ is the broadening factor, $\gamma = 1/\tau + 2/T + 2\pi\delta\nu_{laser}$ is the homogeneous linewidth of the transition in rad/sec, and $I_{sat} = \frac{h\nu_0}{sigma_0\tau}(1+\frac{\tau}{T})(2+\Gamma_i T)^{-1}$ is a convenient yardstick to measure the intensity of the pump beam. In this expression ν_0 is the transition frequency between a hyperfine level of the $5^2 S_{1/2}$ ground state and the i-th hyperfine level of the excited $5^2 P_{3/2}$ manifold, τ is the lifetime of the $5^2 P_{3/2}$ state, Γ_i represents the sum over all spontaneous decay rates from the i-th excited hyperfine level to all lower levels except the initial one, and T is the duration of the excitation (the time that the atoms remain in the laser beam).

The absorption cross section per atom is

$$\sigma_0 = \frac{16\pi^2 k\mu^2}{h\gamma} \frac{1}{4\pi\epsilon_0} \qquad (5.6)$$

where $k = \omega/c$, the electric dipole moment μ satisfies

$$\mu^2 = \frac{2\epsilon_0 hc^3 \Gamma_0}{2\omega_0^3} \qquad (5.7)$$

and Γ_0 is the spontaneous decay rate peculiar to the resonant pair of hyperfine levels. The values of Γ_i and Γ_0 reflect hyperfine branching ratios and satisfy the equation $\Gamma_i + \Gamma_0 = 1/\tau$. Through a given expressions it is evident that greater pump intensities I_{pump} imply stronger but broader signals, and that large values of Γ_i imply small values of I_{sat}, which means that I_{pump} need not be large to produce sizeable saturated absorption signals in the presence of optical pumping. Finally, if T is large compared to τ then in the limit of low laser power, the saturated absorption linewidth Δ is essentially $\delta\nu_{laser}$.

Since the minimum observable linewidth is ~ 6 MHz, the residual measured hyperfine transition ($5S_{1/2}, F=3 \longrightarrow 5P_{3/2}, F=4$) broadening to ~ 10 MHz can be attributed to other broadening effects, mentioned above and also to residual Doppler effect due to a small, but still non-vanishing angle of superimposed pump and probe beams [Dem92].

The final spectrum reveals not only the Doppler free peaks that were discussed, but also additional spurious peaks, appearing between each real absorption peak. These spurious peaks are known as cross-over peaks. These can arise when there is more than one hyperfine transition under the same Doppler profile, allowing the laser beam the possibility of interacting with two different velocity classes of atoms at the same time. Crossover peaks occur because the Doppler shift allows certain moving atoms to be in resonance with both the pump beam and the probe beam. For instance, if the laser is tuned halfway between the transitions from $F = 3$ to $F' = 3$ and from $F = 3$ to $F' = 4$, atoms moving towards the probe beam could see the pump beam as the lower frequency resonance, causing them to depopulate the $F = 3$ ground state, and they could see the probe beam as the higher frequency resonance, resulting in a signal where none ought exist at the frequencies halfway between successive Doppler free transitions.

In the Figure 5.7(a) are depicted three real hyperfine transitions $5S_{1/2}$, $F = 3 \longrightarrow 5P_{3/2}, F = 2, 3, 4$ and the three corresponding crossover peaks, appearing at the halfway frequencies of the transitions $5S_{1/2}, F = 3 \longrightarrow 5P_{3/2}, F = 2$ and $F = 3$; $5S_{1/2}, F = 3 \longrightarrow 5P_{3/2}, F = 2$ and $F = 4$; $5S_{1/2}, F = 3 \longrightarrow 5P_{3/2}, F = 3$ and $F = 4$.

The first stage laser frequency, as already described in the Section 5.2 is stabilized on the required $5S_{1/2}, F = 3 \longrightarrow 5P_{3/2}, F = 4$ hyperfine transition by means of lock-in technique where the feedback is processed by PID regulator.

5.4 Second Stage

The second stage laser ($\lambda_2 = 775, 978$ nm) is used to promote atoms further the ladder to the $5D_{5/2}$ state. To lock the laser on this transition, special absorption spectroscopy setup is realized, where the laser is stabilized on the transmission signal from the first laser. The sketch of such a setup shown in Figure 5.8.

The second stage laser beam, which should be frequency stabilized is superimposed with the light from already frequency locked counter propagating first laser beam in the room temperature Rb vapor cell. The cell is about 10 cm long and enclosed by a μ-metal magnetic shield like in the saturation spectroscopy setup. By using the photodiode (PD), the transmission of the first stage laser is recorded, as the frequency of the second stage is scanned. Since the first stage laser is already locked to the Doppler-free peak, it also excites the atoms in this cell that have no velocity component in the direction of laser beam . If the second stage laser frequency is on resonance ($5P_{3/2}, F = 4 \longrightarrow 5D_{5/2}$), it will deplete the population of the $5P_{3/2}, F = 4$ level and promote partially the atoms to the $5D_{5/2}$ level. Therefore, the portion of atoms that can absorb the light of the first stage laser is reduced, which causes the transition of this laser beam to go up. Due to the different lifetimes of the $5P_{3/2}, F = 4$ and $5D_{5/2}$ states, the stronger signal is obtained from the recorded transmission of the first stage laser beam.

Simple rate model for such system can be estimated using by Einstein introduced A and B coefficients, which describe spontaneous emission and induced absorption and emission. The Einstein A coefficient is defined in terms of the total rate of spontaneous emission W^s_{ik} from an one level $|i\rangle$ to another level $|k\rangle$ for a system of N_i atoms:

$$W^s_{ik} = A_{ik} N_i, \qquad (5.8)$$

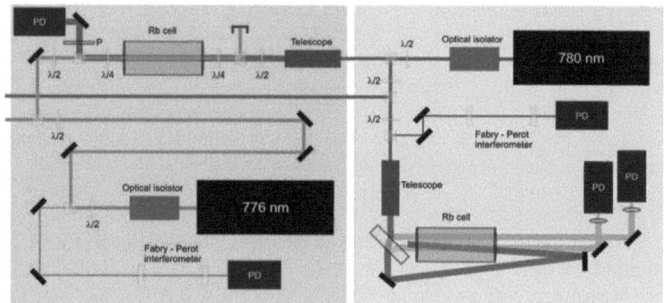

Figure 5.8: Setup of absorption spectroscopy.

where

$$1/t_{spont} = \sum_k A_{ik}. \tag{5.9}$$

The spontaneous emission coefficient A_{ik} is proportional to the Einstein coefficient of induced absorption B_{ki}:

$$B_{ki}^\omega = \frac{g_i}{g_k} B_{ik}^\omega \quad ; \quad A_{ki} = \frac{\hbar \omega_{ik}^3}{\pi^2 c^3} B_{ik}^\omega, \tag{5.10}$$

where ω_{ik} is the resonance frequency of the transition, g_i and g_k are the degeneracy factors of the levels [Hil82].

Sine the lifetime of $5D_{5/2}$ hyperfine levels is approximately ten times longer as the lifetime of the $5P_{3/2}, F = 4$ hyperfine levels (238, 5 ns [SGO08] and 25, 69 ns [The84] respectively), the transmission signal of the first laser beam is more apparent compared to the absorption of the second stage beam.

The excitation scheme used is a bit similar to the electron-shelving [Deh82] technique often used in the trapped ion experiments, called V-excitation scheme. In this study the excitation is different. A ladder-excitation scheme is utilized here.

Since the linewidth of the $5P_{3/2}, F = 4$ transition in this experiment is about 10 MHz (FWHM) as explained before, the hyperfine structure levels $F = 3, 4, 5$ of the $5D_{5/2}$ state can not be simply resolved using usual excitation scheme, because the frequency between them lies in the same range ($\sim$ 9 MHz [NBFM93][GSVD+95]). Therefore, the transitions from all the three hyperfine structure levels will be existent in the last excitation step ($5D_{5/2}F = 3, 4, 5 \rightarrow 63P_{3/2}$), which would spoil the usual maser operation.

5.4. SECOND STAGE

To avoid this, a special excitation scheme is realized where hyperfine transitions $5P_{3/2}, F = 4 \longrightarrow 5D_{5/3}, F = 3, 4, 5$ can be resolved (Figure 5.9a)) and one necessary hyperfine transition can be selected (Figure 5.9b)). For this purpose the optical pumping effect and dipole selection rules are used.

In optical pumping with circularly polarized light [Kas50], the transfer of angular momentum from the light beam to the atomic system results in orientation of the atoms. For circularly polarized photon σ_+ the transition selection rules are $\Delta m_f = +1$. Thus every absorption an excited atom produces one unit more of projected angular momentum than it had before the transition. Spontaneous downward transitions can occur with $\Delta m_f = \pm 1, 0$.

In our setup using circular σ_+ light for both laser beams in the Rb cell and due to the atomic selection rules we realize

$$5S_{1/2}, F = 3, m_f = 3 \to 5P_{3/2}, F = 4, m_f = 4 \to 5D_{5/3}, F = 5, m_f = 5 \tag{5.11}$$

excitation scheme, which is schematically depicted in Figure 5.8. The two laser beams from the first and the second stage (both circularly polarized in the same direction with the help of $\lambda/4$ plates) are superimposed from different sides in the Rb cell. In Figure 5.9 (a) it is shown, that all the three hyperfine transitions are resolved, when the polarizations of the two laser beams were not perfectly circular. When the polarization of the laser beams is set correctly to the circular one, the atoms are optically pumped only into the one hyperfine transition state (Figure 5.9 (b)). The linewidth of the transition is about 15 MHz. Even a smaller linewidth could be obtained in such setup, however the laser powers were optimized to guarantee the best signal-to-noise ratio, even if the linewidth at these power levels is a bit larger than the optimal values.

Anyway, using such a approach the resolved line has much better contrast compared to the methods where the fluorescence signal from the atomic decay $6P_{3/2}$ to the ground state $(420, 3\,\text{nm})$ for the laser frequency stabilization was used [Lan94, Ger08].

The frequency stabilization of the second laser on the required $5P_{3/2}, F = 4, \to 5D_{5/3}, F = 5$, transition is done by means of usual lock-in technique. The feedback signal is processed by PID regulator, similarly like in the first stage frequency locking, described earlier in Section 5.2.

Utilization of selective atomic excitation approach, described here, has led to other experiments, where we have successfully demonstrated the purely optical spectroscopy of rubidium Rydberg states (specifically Rb $63P_{3/2}$) in a room-temperature gas cell. This work of using different polarization schemes and possible excitation geometries with the experimental applications and results are published in [THS+09] and [TGH+09].

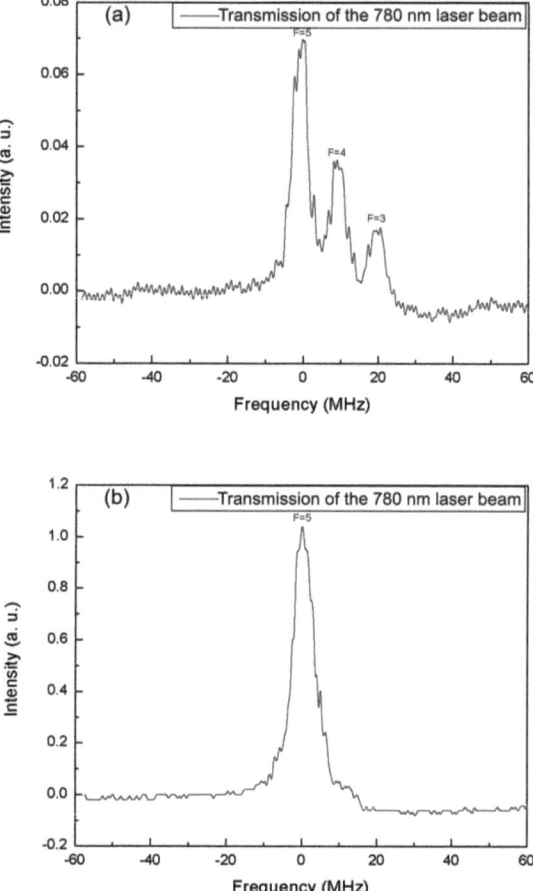

Figure 5.9: Spectroscopy signal for the 2−nd stage laser locking. a) Three hyperfine transitions $5P_{3/2}, F = 4 \rightarrow 5D_{5/3}, F = 3, 4, 5$ can be resolved. b) A necessary hyperfine transition $5P_{3/2}, F = 4 \rightarrow 5D_{5/3}, F = 5$ for the second stage laser stabilization is selected using the effect of optical pumping and dipole selection rules.

5.5 Third Stage

The last laser stage ($\lambda_3 = 1256,730$ nm) is used to promote the Rb atoms from the $5D_{5/3}, F = 5$ state directly into the Rydberg regime ($63p3/2$ state). Since the mean lifetime of the Rydberg states are much longer compared to the lower lying levels used in the first two excitation stages due to the weak overlap of the atomic wave function with that of ground state [FH83], the stabilization of this laser frequency is done directly on the atomic beam signal (5.10) in the auxiliary chamber (see Section 3.4). All the three laser stage beams are superimposed perpendicularly to the atomic beam realizing $5S_{1/2}, F = 3 \rightarrow 5P_{3/2}, F = 4 \rightarrow 5D_{5/3}, F = 5 \rightarrow 63P_{3/2}$ excitation scheme. The atoms in the Rydberg state are detected using field-ionization detection setup.

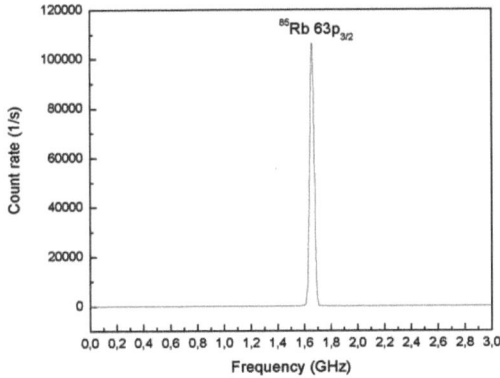

Figure 5.10: The signal from the atomic beam.

Since the obtained signal from the detection system is digital one, stabilization in this case should be done in a bit different way compared to the first two stages, where analog signal from photodiodes was taken. While the usual commercial lock-in amplifiers cannot produce an error signal from the digital input signals, a special circuit was developed for this purpose, which works at the synchronous demodulation principle. The error signal from this circuit afterwards is processed by PI regulator. Such stabilization setup allows reliably to stabilize the frequency of the third laser stage on the peak of the required transition.

For the frequency stabilization quality estimation, atomic beam count rate stability test was performed. In this study all the three lasers were locked and during the fixed interval of time the atomic count rate was recorded (Figure 5.11(a)). In the analysis of the statistics of such measurement, a conclusion about the laser frequency stabilization can be done.

In the ideal case, considering that the laser system works perfectly, frequency is stable, and not only of the last laser stage, but also of the first two, considering that there are no fluctuations in the atomic beam and that the detection system works perfectly, one would expect to get Poisson distribution.

The Poisson distribution is a discrete probability distribution that expresses the probability of a number of events occurring in a fixed period of time if these events occur with a known average rate and independently of the time since the last event. If the expected number of occurrences in the fixed time interval is λ, then the probability that there are exactly k occurrences ($k = 0, 1, 2, ...$) is equal to

$$f(k; \lambda) = \frac{\lambda^k e^{-\lambda}}{k!} \tag{5.12}$$

where k is the number of occurrences of an event - the probability of which is given by the function and λ is a positive real number, equal to the expected number of occurrences that occur during the given interval. The standard deviation of the Poisson distribution is $\sqrt{\lambda}$.

Such experimental results are plotted in Figure 5.11(b) together with the theoretical Poisson curve. Experimental statistics is very close to the fundamental physical limit(within 6 %), which confirms the high laser stabilization quality (for all three stages) and overall stability of the system.

5.5. THIRD STAGE

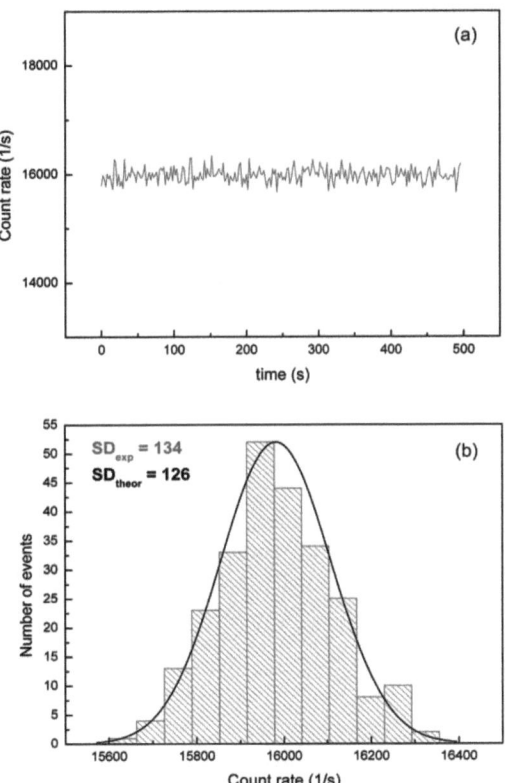

Figure 5.11: The test of the laser stabilization.

Chapter 6
Magnetic Field Compensation

6.1 Overview

Without compensation in the experimental region, existent magnetic field is of the order of 350 mG (mainly earth magnetic field). Since the resonator is made from niobium, which is superconductor of type II, below the superconductive critical temperature (9.2 K) the magnetic field fluxes freeze inside. This causes increased high frequency resistance [Pie73] in the walls of resonator, worsens it's Q-factor [PSS93] and reduces the lifetime τ_{cav} of the field.

Another reason for compensation is that in magnetic field magnetic sublevels of the transitions, used in the maser experiments, are no longer degenerate. Due to the different Lande-factors the transitions $63P_{3/2}, m_j = 1/2 \rightarrow 61D_{5/2}, m_j = 1/2$ and $63P_{3/2}, m_j = -1/2 \rightarrow 61D_{5/2}, m_j = -1/2$ splits about 1.5 kHz/mG. Since the splitting should be smaller than Rabi frequency $\Omega/2\pi$, the magnetic field should be compensated to at least 2 mG [Rai95].

Rough magnetic field precompensation down to ~ 30 mG is done using magnetometer (Magnetoscop type 1.068)

Due to experimental reasons, the final precise magnetic field compensation (MFC) in maser is possible when atoms themselves are used as the detectors. For this purpose, the precession of magnetically oriented atoms in a variable magnetic field is observed.

The main setup of a magnetic field compensation is shown in Figure 6.1. Outside the cryostat, three pairs of perpendicular Helmholtz coils (diameter ~ 1 meter) are arranged in a way that the resonator lies in the center point. The current through each pair of the coils is controlled by high-precision current supplies (Burster Digistant 6426). With the help of the first laser, the magnetic moments of all atoms are aligned along the direction of this

92   CHAPTER 6. MAGNETIC FIELD COMPENSATION

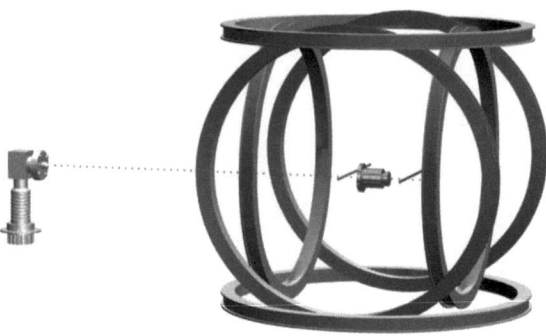

Figure 6.1: The main setup of magnetic field compensation in one-atom maser.

laser beam, in this way polarizing the atomic sample. Then the atoms move through the region of about 10 cm where magnetic field should be compensated, namely, resonator. In this region the magnetic moment of the atoms precesses around non vanishing magnetic field. Afterwards, such a precession (Hanle effect [2] [Han24]) is detected by subsequent excitation of the atoms to the Rydberg levels with a second laser.

A more detailed explanation is shown in Figure 6.2. Atoms from the oven with the thermally equally distributed quantum numbers using the circular light of the 780 nm diode laser are pumped to the extremal quantum state:

$$5^2S_{1/2}, F = 3 \rightarrow 5^2P_{3/2}, F = 4 \qquad (6.1)$$

Since only the transitions of $\Delta m_f = +1$ are allowed (laser beam direction defines the quantization axis), after several hundred of absorption and spontaneous emission cycles in the laser beam (laser beam diameter $\sim$ 3 mm; average velocity of atoms $\sim$ 300 m/s and the lifetime of excited state 26 ns gives $\sim$ 380 cycles) the atoms leave this excitation region in a $5^2S_{1/2}, F = 3, m_f = 3$ state with a well aligned magnetic moment.

So if the atoms are moving along a certain distance, where they are subjected to a static homogeneous magnetic field, then during the time of

―――――――――――――
[2]Whereas in the original experiments the rotation of the polarization of emitted fluorescence of an atomic ensemble was examined, the name Hanle-effect is now used for a variety of experiments which involve an optical detection of rotating atomic magnetic moments.

6.1. OVERVIEW

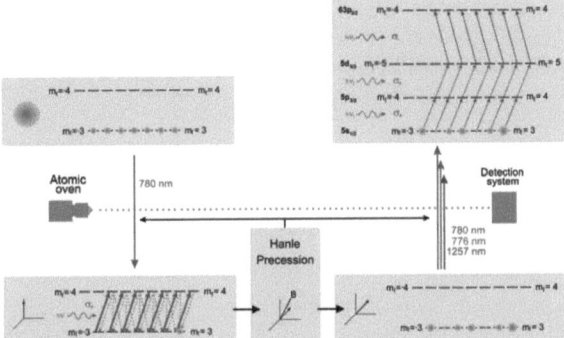

Figure 6.2: Quantum mechanical explanation of magnetic field compensation in one-atom maser.

passage, the hyperfine structure spin will precess around magnetic field, thereby changing the distribution of m_f quantum number states in a coherent manner. The probability that the spin after the time t will be oriented again in the initial direction ($m_f = +3$) is given by

$$P_{m_f=+3}(t) = 1 - \sin^2\alpha \sin^2(\omega t) \tag{6.2}$$

where α is the angle between magnetic field and laser beam and $\omega = g_F \frac{eB}{2m}$ is Larmor frequency. The constant g_F is hyperfine structure Lande-factor (here 1/3). This theoretical model comes from the calculations done by investigating the coherent atomic control of the atomic states described in Section 6.3.

At the end of the region atoms are promoted to the Rydberg state with the three-step laser system which was already described in Chapter 5. The polarization of three laser beams are chosen in a way (corresponding to $\sigma_+, \sigma_+, \sigma_-$) that in 0 magnetic field the excitation rate over the selected ladder

$$5S_{1/2}, F = 3, m_f = +3 \rightarrow 5P_{3/2}, F = 4, m_f = +4 \rightarrow$$
$$5D_{5/2}, F = 5, m_f = +5 \rightarrow 63P_{3/2}, F = 4, m_f = +4 \tag{6.3}$$

is most efficient (i.e. the highest excitation probability is achieved in vanishing magnetic field, when there is no precession of the magnetic moment). However in the presence of magnetic field, the excitation probability of atoms depends on m_f quantum number. Depending on how far the spin has precessed, the resulting superposition of m_f quantum levels will be different.

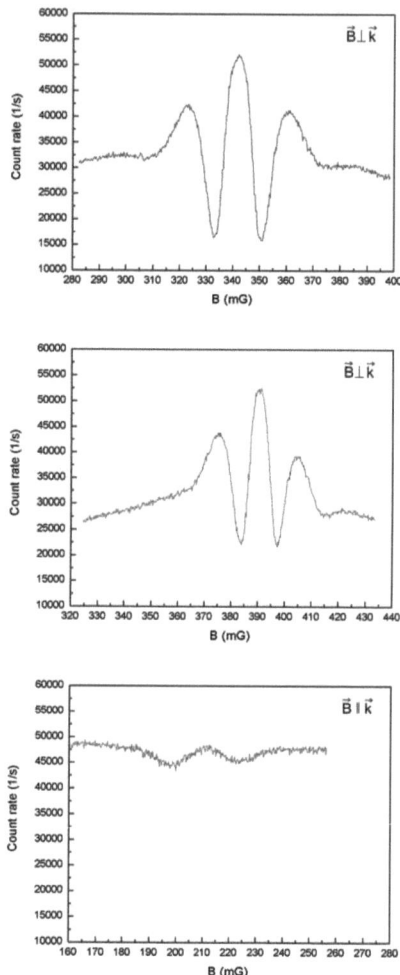

Figure 6.3: The experimental results of magnetic field compensation in one-atom maser for all the three spatial directions.

Because each m_f transition has characteristic from Clebsch-Gordan coefficients dependent oscillator strength, the probability at which the atom will be promoted into the Rydberg regime depends on the direction and magnitude of magnetic field and the interaction time.

Since the precession of a spin is a periodic process, the excitation rate for an atoms with a fixed velocity depends on magnitude of magnetic field periodically also (oscillatory behavior of the count rate excitation to the Rydberg states). Each of these oscillations corresponds to complete rotation of a spin. However, since in the usual setup the atoms with all velocity components are excited with the laser that is perpendicular to the atomic beam, higher order oscillations are smeared out. Only at $B = 0$ field - the symmetry center of the curve - the oscillation has a pronounced extremum since the magnetic moment of all atoms remains temporally aligned. The required magnetic field compensation point lies in the symmetry center of this curve.

Such magnetic field compensation using an iterative method is done in all three directions. There is one special case, when magnetic field is in the direction of quantization axis, namely laser beam ($\vec{B}\|\vec{k}$) direction. In this case the magnetic moment of atoms and the magnetic field is collinear and the spin precession around itself yields a zero net result. The magnetic field compensation in this direction is done by applying a small ($\sim 3\,\mathrm{mG}$) magnetic field offset in one of the other directions.

The experimental results of magnetic field compensation are shown in Figure 6.3. Here the final iterative scans for each of three directions are shown. The right compensating magnetic field is determined from the symmetry center of the curve. The magnetic field compensation in this case is done with the precision of about 1 mG.

Such magnetic field compensation measurements have been routinely performed in one-atom-maser experiments before as well, but increased reliability, stability and efficiency of the system allows to compensate magnetic field better, compared to those measurements [Bod00, Bra01], which have been performed previously.

6.2 High Resolution Magnetic Field Compensation

For even more precise magnetic field compensation, a new method was developed and realized in the experiment. The resolution with which magnetic field is compensated, using this new modulation-demodulation technique dramatically increased by several orders of magnitude. Using such a method magnetic field compensation was performed for the first time in the one-atom-maser experiments.

This method employs a lock-in based detection scheme which is used to

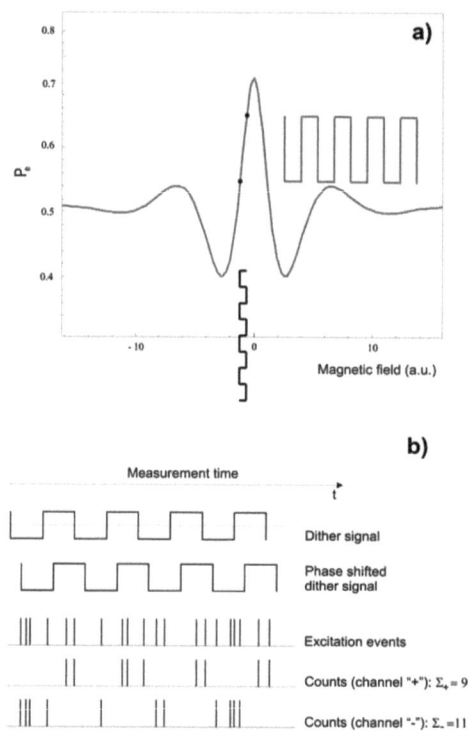

Figure 6.4: The modulation - demodulation scheme for the magnetic field compensation.

6.2. HIGH RESOLUTION MAGNETIC FIELD COMPENSATION 97

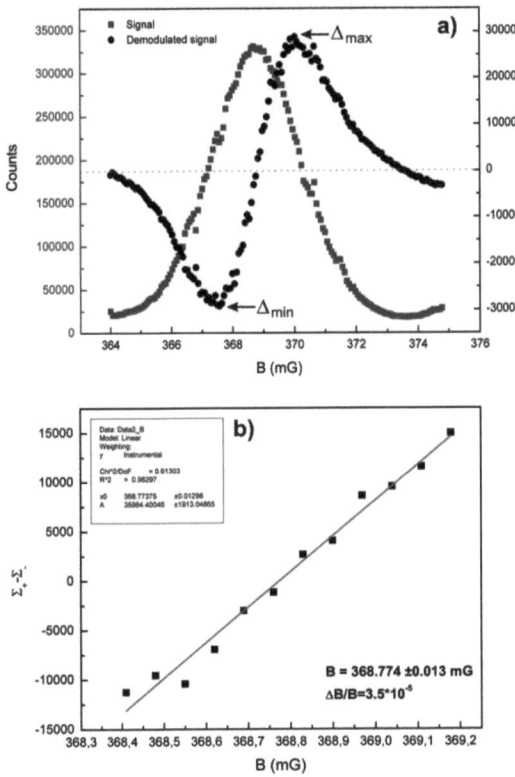

Figure 6.5: Experimental results of the modulation - demodulation magnetic field compensation.

determine the compensating coil current with strongly enhanced resolution. The main idea is graphically represented in Figure 6.4. The basic steps are:

- Small rectangular modulation ("dither") is superimposed on the DC-current ("offset") of one of the Helmholtz pairs. For the realization of this new method on each of the Helmholtz coils an additional 10 wire-windings (applied 1 mA current correspond to $B = 0.225$ mG) for the field modulation were added. Modulation frequency was limited by the induced induction in the coils and the optimum value of 2 Hz was chosen.

- A phase shifted signal defines two detection channels: channel "+" and channel "-". Channel "+" ("-") is open if and only if the phase shifted dither signal has a positive (negative) polarity.

- Separate counters determine the excitation events for each of the two channels. The demodulation signal is obtained by subtraction.

Such measurement is performed for a set of offset currents that are sufficiently dense in the vicinity of the compensating current. Figure 6.5(a) shows the experimental results. The blue points correspond to usual magnetic field compensation signal as it was described in the previous section. The black points corresponds to the dispersion signal. A linear fit $y = a(B - B_0)$ (Figure 6.5(b)) to the demodulated signal based on all measured peaks with:

$$\frac{\Delta_{min}}{2} \leq \Delta \leq \frac{\Delta_{max}}{2} \qquad (6.4)$$

yields the compensating field B_0 and its error, that gives an absolute resolution of ~ 10 µG.

Such high resolution magnetic field determination and compensation in the micromaser experiments was achieved for the first time. The obtained result is three orders of magnitude better compared to those measurements with "several mG" resolution which have been performed previously [Rai95][Bra01].

6.3 Velocity Selected Magnetic Field Compensation

The new three-step-excitation setup opened the possibility for the detail investigation of the atomic Hanle precession in the static homogenous magnetic field. It also made a magnetic field compensation measurements with a velocity selected atoms possible. Such kind of measurements were done for the first time in one atom maser experiments. Performing such measurements allows not only to compensate magnetic field with a higher resolution compared to the measurements with a thermal atomic beam (see Section

6.3. VELOCITY SELECTED MFC

6.1), but also to perform quantum coherent control of the magnetic substate of the atoms.

The setup of the velocity selected magnetic field compensation (MFC) is similar to the one used in MFC experiments with the thermal atomic beam (see Section 6.1). However for the certain atomic velocity group selection via the Doppler effect a different excitation scheme to the Rydberg states was utilized.

The main principle of the experiment depicted in Figure 6.6. With a circularly polarized light of the first laser ($\lambda = 780$ nm), the magnetic moment of all atoms are polarized along the laser direction by means of optical pumping. The laser beam direction ($\vec{k}$) is also chosen as the quantization axis. After several hundreds of absorption and spontaneous emission cycles in the laser beam, the atoms leave this excitation region in a $5S_{1/2}, F = 3, m_f = 3$ state with a well aligned magnetic moment. Afterwards atoms move along a certain distance (around 8 cm) including the resonator where they are subjected to a static homogeneous magnetic field.

Magnetic field is produced by three pairs of Helmholtz coils which were already described in Section 6.1.

During the time of passage the magnetic moment of atoms precess around magnetic field thereby by changing the distribution of m_f quantum number states in a coherent manner (Hanle precession [Han24]).

At the end of a region using diode lasers (described in Chapter 5) in a three-step-ladder ($\lambda_1 = 780$ nm; $\lambda_2 = 776$ nm; $\lambda_3 = 1256$ nm) atoms are promoted to the Rydberg state $63P_{3/2}$ and detected using a state-selective field-ionization detection system.

To realize the excitation scheme depicted in Figure 6.6, the polarizations of the laser beams are chosen correspondingly. For this purpose the first two laser stages are circularly (σ^+) polarized and superimposed to hit the atomic beam perpendicularly. The third laser beam is placed along the atomic beam and is linearly (π) polarized along the propagation direction of the first and second stage lasers. This laser is also used for the selection of the certain well defined velocity group of the atoms by means of the Doppler effect.

The measurements were done at 77 K temperature. The experimental results of such a Hanle precession are shown in Figure 6.7. The detection of atoms in the Rydberg state is measured as the magnetic field in the direction perpendicular to the pump laser beam ($B \perp \vec{k}$) is scanned. Here the atoms with the 345 m/s velocity were chosen.

The laser power were: 5 mW for optical pumping, $P_{\lambda_1} = 5$ μW, $P_{\lambda_2} = 4$ mW and $P_{\lambda_3} = 10$ mW. The power of the λ_1 laser were chosen correspondingly low, to avoid optical pumping and redistribution of atoms where Hanle precession would be hardly observable (see Figure 6.10).

The symmetry center of the curve shows the compensated ($B = 0$) earth and ambient magnetic field. In this scan direction ($B \perp \vec{k}$) it is at $B =$

100   CHAPTER 6. MAGNETIC FIELD COMPENSATION

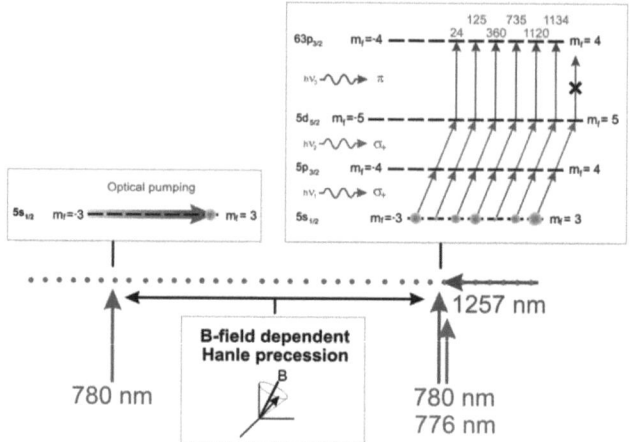

Figure 6.6: The scheme of experimental coherent atomic control realization. With the first laser, the magnetic moments of atoms are aligned by means of optical pumping. Then the atoms pass through the region where they are subjected to a static magnetic field. At the end of the region, the atoms are promoted to the Rydberg regime by a three-step excitation. Depending how far the spin has precessed in the magnetic field, the resulting m_f quantum number state distribution will be different. This is observed by detection of atoms in the Rydberg regime, because every m_f have different oscillator strength, therefore different excitation probability (noted in numbers above each excitation path).

342 mG. In the other $B \perp \vec{k}$ direction the compensation is done in similar way.

However, the case where the magnetic field is applied in the direction of quantization axis ($B \parallel \vec{k}$), should be discussed separately. The precession of atomic magnetic moment in this case can not be observed due to 0 deg angle between the induced magnetic moment and the magnetic field. To make the Hanle precession effect visible, a small (~ 2 mG) magnetic field offset in one of the other directions should be applied. When one applies such a small offset in the one of the perpendicular directions, the magnetic moment of atom will precess around this component. However, when the B-field in the laser direction is strong, this precession would be hardly observable again, because the effect of this applied component would be in comparison very small. So only when the B-field in the laser direction is small then this small offset starts to play a dominant role and the precession can be observed. Such measurement is shown in Figure 6.8. That is why in this scan direction

6.3. VELOCITY SELECTED MFC

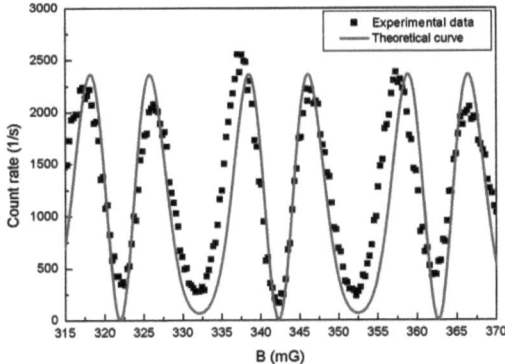

Figure 6.7: Theoretical fit and experimental results.

the oscillations are observable only at almost compensated magnetic field (here compensated magnetic field corresponds to $B = 210$ mG).

Such magnetic field compensation measurement using coherent control of atoms is done in all the three directions in an iterative way.

Figure 6.9 shows a scan for different velocity groups of atoms. Since the different velocities of atoms takes a different time of passage through the region where magnetic field is applied, the accumulated precession angle is different also. Therefore the period of the Hanle precession is smaller for the higher velocities of atoms. Due to the different number of atoms in each velocity group the corresponding count rate is different also. For the clarity reasons the count rate here was normalized to the symmetry point of the curve. At the compensated magnetic field for all the atoms no precession occurs what leads to the dark state. This is clearly pbserved from the point where all the curves overlap ($B_0 = 342$ mG). From such measurements, the magnetic field compensation with the precision of the 0.3 mG in each of the three directions is done. This gives more than 3 times better resolution than using common (see Section 6.1) magnetic field compensation approach, but is less sensitive than demodulation technique (described in Section 6.2) where the absolute resolution of ~ 10 μG is achieved. Nevertheless, such measurements were performed not to increase the B-field compensation resolution even more, but for better investigation and understanding of coherent control effect of atoms.

As mentioned above, if in the excitation to the Rydberg states process

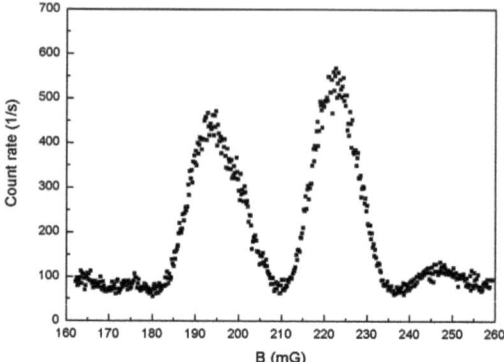

Figure 6.8: Experimental results of magnetic field scan in the $B \parallel \vec{k}$ direction.

the laser power of the first stage is sufficient to make an optical pumping, it affects the final Hanle precession signal. It is shown in Figure 6.10, where the power of the first stage laser was $P_{\lambda_1} = 6.2\,\text{mW}$, as compared to the $P_{\lambda_1} = 5\,\mu\text{W}$ used in measurements shown in Figure 6.7. In this case the secondary minimas are suppressed due this optical pumping effect. The qualitative explanation can be done by investigating the whole excitation to the Rydberg states scheme. If one has a $\sigma_+\sigma_+\pi$ scheme, the excitation probability in the Rydberg states grows towards the direction of extremal m_f state (see the numbers over each of excitation paths in Figure 6.6). If there is not dark state at the extremal $m_f = 5$, one would observe a maximum count rate for a perfectly compensated magnetic field. But in addition to this behavior, one has an abrupt cut, because the excitation to the extremal $63P_{3/2}\,m_f = 5$ state probability is zero, which gives a qualitatively sharper minimum for the compensated magnetic field (see Figure 6.7). Therefore, one has a double structure in this curve.

Now if the sufficient optical pumping takes place in the excitation path to the Rydberg states, the population towards the extremal m_f states occurs as in the case for the compensated magnetic field. Therefore the case where $B = 0$ is not affected. However, in this case the secondary minima (which is the effect of the negative m_f sublevels, which population is depleted by optical pumping) is suppressed (Figure 6.10).

To explain the observed data the optical pumping, the evolution of the atoms in a magnetic field and the detection process have to be described

6.3. VELOCITY SELECTED MFC

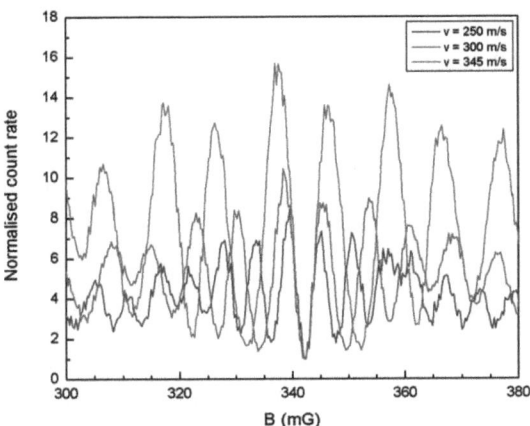

Figure 6.9: Experimental results of magnetic field scan for different classes of atomic velocities.

separately.

The calculation approach chosen here is formally very simple (similar to the one described in [BKD08] where the Hanle effect calculation is done for the simple $F = 0 \to F = 1$ transition). It has the advantage that the influence of different processes can be followed directly. The main idea is represented in Figure 6.11:

1. Starting with the ideal case, it is assumed that the optical pumping is ideal and laser beam direction is chosen as the quantization axis, where the optical pumping is described as σ_+.

2. Since chosen atomic coordinate system does not coincide with the B-field direction, in order to describe precession, a coordinate rotation using the Wigner D-matrix into the direction of the magnetic field is necessary, where the magnetic field is chosen now as the quantization axis (and is perpendicular to laser direction). One has to calculate the Wigner D-matrix for arbitrary rotation angle ($\alpha = 0$, $\beta = 0$, $\gamma = \pi/2$; here α, β and γ are the Euler angles) in 7 dimensions (for $5s_{1/2}, F = 3$

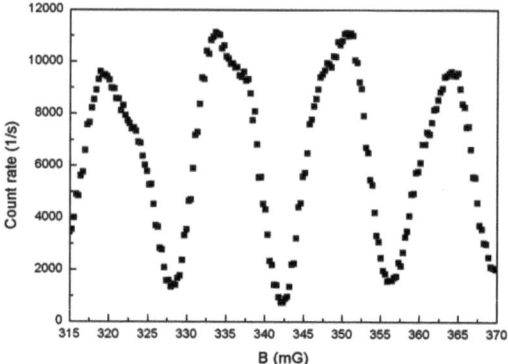

Figure 6.10: The scan of magnetic field with the higher first stage laser power $P_{\lambda_1} = 6.2$ mW due to the optical pumping effect shows spoiled Hanle precession.

and each m_f state) according to the [Wei78]:

$$D^{(j)}_{m'm}(\alpha\beta\gamma) = \sum_\lambda \frac{(-1)^\lambda \sqrt{(j+m)!(j-m)!(j+m')!(j-m')!}}{\lambda!(j+m-\lambda)!(j-m'-\lambda)!(m'-m+\lambda)!} \\ \times e^{im'\alpha}(\cos\frac{1}{2}\beta)^{2j+m-m'-2\lambda}(\sin\frac{1}{2}\beta)^{m'-m+2\lambda}e^{im\gamma} \quad (6.5)$$

3. The interaction with the magnetic field. In this case the phase factor $e^{im_f\omega_L t}$ for each m_f state is added, where ω_L is the Larmor frequency and t is the interaction time.

4. Wigner D-matrix is rotated back, where the coordinate system again is chosen in the detection laser beam direction.

5. Finally, the excitation transition probabilities are calculated by using general formula [MvdS99] for each transition separately (see Figure 6.6):

$$\mu_{eg} = e(-1)^{1+L'+S+J+J'+I-M'_F}\langle\alpha'L'||r||\alpha L\rangle \\ \times \sqrt{(2J+1)(2J'+1)(2F+1)(2F'+1)} \\ \times \begin{Bmatrix} L' & J' & S \\ J & L & 1 \end{Bmatrix} \begin{Bmatrix} J' & F' & I \\ F & J & 1 \end{Bmatrix} \begin{pmatrix} F & 1 & F' \\ m_f & q & -m_f \end{pmatrix} \quad (6.6)$$

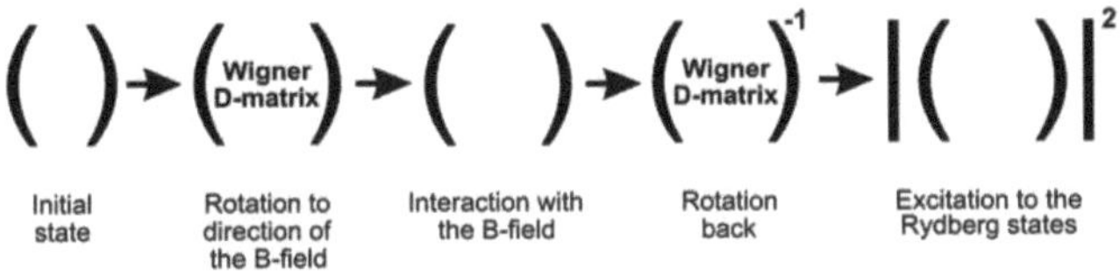

6.3. VELOCITY SELECTED MFC

$$\left(\ \right) \rightarrow \boxed{\substack{\text{Wigner}\\ \text{D-matrix}}} \rightarrow \left(\ \right) \rightarrow \boxed{\substack{\text{Wigner}\\ \text{D-matrix}}}^{-1} \rightarrow \left|\left(\ \right)\right|^{2}$$

| Initial state | Rotation to direction of the B-field | Interaction with the B-field | Rotation back | Excitation to the Rydberg states |

Figure 6.11: Scheme of the theoretical model for the Hanle precession calculation.

Here the $3 - j$ and $6 - j$ symbols were taken from [RBMJKW59]. In the calculations it is assumed, that the magnetic field is constant over the whole interaction region. Asymmetries in the observed spectra can be attributed to such imperfections and therefore are not reproduced by the calculations.

For easier and faster calculation and variation of the fitted experimental parameters the whole theoretical model process was programmed.

For the explanation of the observed signals when the magnetic field is scanned over a wide range (Figure 6.12), two additional effects have to be taken into account:

- First, optical pumping in a finite magnetic field is not perfect and the atoms no longer start their interaction with the magnetic field in the extremal m_f state.

- Second, the averaging due to a non perfect velocity selection has to be taken into account.

While the first effect can be treated by including a magnetic precession into the description of the optical pumping process, the second effect is treated by averaging several obtained spectra over different velocities. The velocity spread of the atoms can be calculated in a time of flight setup and was determined to be 8% (see Section 4.3). Figure 6.12 shows that all relevant features of the measurement can be reproduced in the theoretical calculation.

The theoretical fit to the measured curve in Figure 6.7 was done by using analytical approach. First, it was assumed that all atoms are pumped by the 780 nm laser into the extremal m_f state and start interacting with the magnetic field in the $5S_{1/2}$, $F = 3$, $m_f = 3$ ground state. If the direction of the magnetic field does not coincide with the propagation direction of the 780 nm laser, the coordinate system has to be rotated with a 7×7 Wigner D–Matrix. As the state selective field ionization of the Rydberg atoms is insensitive to the m_J quantum number of the Rydberg atom, the observed signal depends on the probability to excite the ground state atom into the Rydberg manifold with the three step laser system. If the magnetic field is perpendicular to the propagation direction of the first two lasers, an analytic

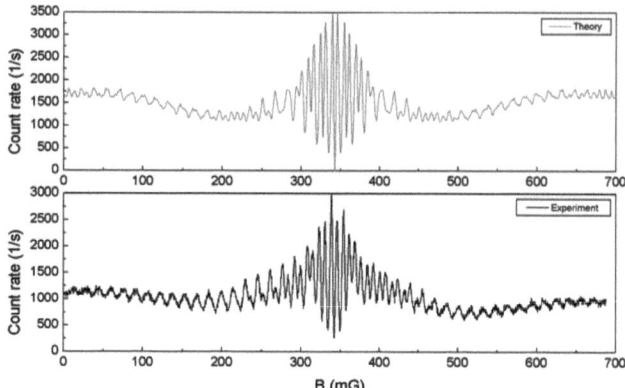

Figure 6.12: Theoretical model and experimental measurement of the Hanle precession over a large range of magnetic field scan.

formula for the excitation probability of the atoms is given as a sum over 6 cosine-terms with real amplitudes A_i, where only angular part of transition matrix element gives an overall scaling and the calculated curve has to be multiplied by a P_0 factor to match the count rate:

$$P = P_0 \cdot (A_0 + \sum_{i=1}^{i=6} A_i \cdot \cos(2i \cdot \omega_L t)) \qquad (6.7)$$

Here ω_L is the Larmor frequency and t the interaction time. If the excitation is performed via the $5D$; $F = 5$ level, dipole selection rules ensure, that only the $63P$, $F = 4$ level can be excited in the Rydberg manifold. So a particular hyperfine level of the Rydberg manifold can be selected even if these levels cannot be individually resolved in spectroscopy. For this case, the prefactors A_i are given in Table 6.1. While in the first two labeled excitation schemes the case of zero magnetic field corresponds to a dark state, it corresponds to an excitation maximum in the case of $\sigma_+ \sigma_+ \sigma_-$ excitation. For comparison, in the last column the excitation probability for a single step excitation from the $5S_{1/2}$ ground state to the $63P_{3/2}$ Rydberg state with σ_+ polarized light is given, where such an excitation scheme was utilized in former magnetic field compensation measurements.

The dependence of this probability on the magnetic quantum number of the ground state is shown in Figure 6.6. As each hyperfine level of the

6.3. VELOCITY SELECTED MFC

Table 6.1: Coefficients in the general expression for the excitation probability for different excitation schemes. All coefficients are expanded only to give integer numbers and have to be scaled to match the overall excitation probability. The last column represents one step excitation used in the previous micromaser experiments.

	$\sigma_+ \sigma_+ \pi$	$\sigma_+ \sigma_+ \sigma_+$	$\sigma_+ \sigma_+ \sigma_-$	σ_+
A_0	131026	4942	2043398	2
A_1	60464	-1912	2993272	1
A_2	-116555	-4625	1163435	0
A_3	-64040	820	229580	0
A_4	-10370	706	20810	0
A_5	-520	68	700	0
A_6	-5	1	5	0

ground state ends up in one specific substate of the Rydberg manifold due to well selected laser polarizations and therefore well defined excitation paths, interferences between different excitation pathways are not observed and not included in the calculation.

Figure 6.7 the shows the excitation probability for slight detuning of the magnetic field around zero. The polarizations of the three lasers are $\sigma_+\sigma_+\pi$. Minima can be observed for Larmor precession angles of 0, π and 2π. Whereas the minima at the even multiples of π can be explained as dark states of the excitation process, the minima at the odd multiples of π are due to the small excitation probabilities for negative m_f states. The sensitivity of magnetic field detection is therefore increased by a factor of two compared to a one step excitation process where minima occur only at odd or even multiples of π.

As mentioned in the beginning, such kind of measurements were done for the first time in the one atom maser experiments. Using the new three step laser excitation scheme opened the possibility to make precise magnetic field compensation by selecting just one velocity group of the atoms.

This method can be used not only to control the ambient magnetic field in an atomic beam apparatus but also to control the magnetic substate and Zeeman coherences of the atoms in a well defined interaction region. Applying a small magnetic field in a specified direction and selecting the velocity on the excited atoms it is possible deterministically control the magnetic substate of the atoms after the interaction region. In principle every magnetic state which corresponds to a magnetic moment pointing in a well defined direction can be produced on demand. This demonstrates the possibility of coherent control of magnetic substates in an atomic beam.

If the setup is seen as an atom interferometer, the rotation of the quantization axis in the direction of the magnetic field puts the atom in a coherent superposition of m_F sublevels, each of which acquires a different phase in the magnetic field. By rotating the quantization axis back into the propagation axis of the probing lasers the 7 different interferometer paths are superposed again, and an interference signal is observed.

Bibliography

[ACDF94] An, Kyungwon, James J. Childs, Ramachandra R. Dasari, and Michael S. Feld: *Microlaser: A laser with one atom in an optical resonator*. Phys. Rev. Lett., 73(25):3375 – 3378, 1994.

[AFH+71] Allen, M. A., Z. D. Farkas, H. A. Hogg, E. W. Hoyt, and P. B. Wilson: *Superconducting niobium cavity measurements at slac*. IEEE Transactions on Nuclear Science, 18(3):168–172, June 1971.

[Ant99] Antesberger, Gunter: *Phasendiffusion und Linienbreite beim Ein-Atom-Maser*. Dissertation, Ludwig-Maximilians-Universität München, 1999.

[Bab89] Babst, G.: *Aufbau und test eines Mikromaser mit supraleitenden Resonator im Temperaturbereich von 0.5 K*. Diplomarbeit, Ludwig-Maximilians-Universität, München, 1989.

[Bal01] Balshaw, N. H.: *Practical Cryogenics*. Oxford Instruments Superconductivity Limited, Old Station Way, Eynsham, Witney, Oxon, OX29 4TL, England, 2001.

[Ben95] Benson, Oliver: *Experimentelle Untersuchung von Bistabilität und Atominterferenz im Ein-Atom-Maser*. Dissertation, Ludwig-Maximilians-Universität München, 1995.

[BK86] Barnett, S. M. and P. L. Knight: *Dissipation in a fundamental model of quantum optical resonance*. Phys. Rev. A, 33(4):2444–2448, April 1986.

[BKD08] Budker, Dmitry, Derek F. Kimball, and David P. DeMille: *Atomic Physics*. Oxford University Press, Great Clarendon Street, Oxford OX2 6DP, second edition, 2008.

[BM98] Buchleitner, Andreas and Rosario N. Mantegna: *Quantum stochastic resonance in a micromaser*. Phys. Rev. Lett., 80(18):3932 – 3935, 1998.

[Bod00]　　Bodendorf, Christof Tilmann: *Untersuchungen am Ein-Atom-Maser mit externer Einkopplung.* Dissertation, Ludwig-Maximilians-Universität München, 2000.

[Bra01]　　Brattke, Simon: *Untersuchung von Photonenzahlzuständen mit dem Ein-Atom-Maser.* Dissertation, Max-Planck-Institut für Quantenoptik, 2001.

[Bur]　　Burle, Industries Inc.: *Channeltron electron multiplier handbook for mass spectrometry applications.* www.burle.com.

[BVW01]　　Brattke, Simon, Benjamin T. H. Varcoe, and Herbert Walther: *Generation of photon number states on demand via cavity quantum electrodynamics.* Phys. Rev. Lett., 86(16):3534 – 3537, 2001.

[CFL+03]　　Casagrande, F., A. Ferraro, A. Lulli, R. Bonifacio, E. Solano, and H. Walther: *How to measure the phase diffusion dynamics in the micromaser.* Phys. Rev. Lett., 90(18), 2003. id. 183601.

[CL05]　　Casagrande, F. and A. Lulli: *Atomic correlations and cavity field decoherence in a strongly driven micromaser.* J. Opt. B, 7:437–444, 2005.

[Deh82]　　Dehmelt, Hans: *Mono-ion oscillator as potential ultimate laser frequency standard.* IEEE Transactions on Instrumentation and Measurement, 31(83), 1982.

[Dem92]　　Demtröder, Wolfgang: *Laser Spectroscopy. Basic Concepts and Instrumentation.* Springer-Verlag GmbH, Berlin Heidelberg New York, 2nd edition, 1992.

[ELS96]　　Elmfors, Per, Benny Lautrup, and Bo Sture Skagerstam: *Dynamics, correlations, and phases of the micromaser.* Phys. Rev. A, 54(6):5171 – 5192, 1996.

[FH83]　　Fabre, C. and S. Haroche: *Rydberg states of atoms and molecules.* Cambridge University Press, Cambridge, 1983. p. 117–164.

[FJM86a]　　Filipowicz, P., J. Javanainen, and P. Meystre: *Quantum and semiclassical steady states of a kicked cavity mode.* J. Opt. Soc. Am. B, 3:906–910, 1986.

[FJM86b]　　Filipowicz, P., J. Javanainen, and P. Meystre: *Theory of a microscopic maser.* Phys. Rev. A, 34(4):3077–3087, October 1986.

BIBLIOGRAPHY

[FSH+05] Florescu, M., S. Scheel, H. Häffner, H. Lee, D. Strekalov, P. L. Knight, and J. P. Dowling: *Single photons on demand from 3d photonic band-gap structures.* Europhys. Lett., 69(6):945–951, 2005.

[FYYH+06] Fang-Yen, C., C. C. Yu, S. Ha, W. Choi, K. An, R. R. Dasari, and M. S. Feld: *Observation of multiple thresholds in the many-atom cavity qed microlaser.* Phys. Rev. A, 73(4), 2006. id. 041802.

[Ger08] Germann, Thomas: *Experimente für den Ein-Atom-Maser mit Hilfe eines 3-stufigen Diodenlaser-Systems.* Diplomarbeit, TU München and MPQ, 2008.

[GSVD+95] Grove, T. T., V. Sanchez-Villicana, B. C. Duncan, S. Maleki, and P. L. Gould: *Two-photon two-color diode laser spectroscopy of the rb $5d_{5/2}$ state.* Physica Scripta, 52:271–276, 1995.

[Han24] Hanle, W.: *Über magnetische Beeinflussung der Polarisation der Resonanzfluoreszenz.* Zeitschr. f. Phys., 30:93, 1924.

[Hil82] Hilborn, Robert C.: *Einstein coefficients, cross sections, f values, dipole moments, and all that.* American Journal of Physics, 50(11):982–986, November 1982. Revised version: February 2002.

[HR06] Haroche, Serge and Jean Michel Raimond: *Exploring the Quantum: Atoms, Cavities, and Photons.* Oxford University Press, 198 Madison Avenue, New York, NY 10016 U.S.A., 2006.

[Jac99] Jackson, J. D.: *Classical Electrodynamics.* John Wiley and Sons, Inc, Toronto, third edition edition, 1999.

[JC63] Jaynes, E.T. and F.W. Cummings: *Comparison of quantum and semiclassical radiation theories with application to the beam maser.* Proceedings of the IEEE, 51(1):89–109, January 1963.

[Kas50] Kastler, Alfred. J. Phys. Radium, 11:255–265, 1950.

[Kle89] Klein, Norbert: *Supraleitende Mikrowellenresonatoren für Anwendungen in physikalischen Grundlagenexperimenten.* Dissertation, Bergische Universität Gesamthochschule Wuppertal, 1989.

112 BIBLIOGRAPHY

[KLZ83] Kleppner, D., Michael G. Littman, and Myron L. Zimmerman: *Rydberg states of atoms and molecules*. Cambridge Univesity Press, The Pitt Biulding, Trumpington Street, Cambridge CB2 1RP, 1983.

[Lan94] Lange, Wolfgang: *Kohärent angeregte Rydberg-Atome in einem resonatormodifizierten Vakuumfeld*. Dissertation, Max-Planck-Institut für Quantenoptik, 1994.

[Lid90] Lide, David R. (editor): *CRC handbook of chemistry and physics*. CRC Press, Boca Raton, Ann Arbor, Boston, 71st edition, 1990.

[LP79] Liberman, S. and J. Pinard: *Experimental studies of highlying rydberg states in atomic rubidium*. Phys. Rev. A, 20(2):507–518, August 1979.

[Lui01] Luiten, Andre N. (editor): *Frequency Measurement and Control*, volume 79 of *Topics in Applied Physics*. Springer, Berlin/Heidelberg, 2001.

[Mar03] Marchi, Gabriele: *Construction of an aparatus for the measurement of the phase diffusion in the micromaser*. Thesis, Universita Degli Studi di Milano, 2003.

[Mes84] Meschede, Dieter: *Strahlungswechselwirkung von Rydbergatomen Realisierung eines Ein-Atom-Masers*. Dissertation, Ludwig-Maximilians-Universität München, 1984.

[MKB+00] Michler, P., A. Kiraz, C. Becher, W. V. Schoenfeld, P. M. Petroff, Lidong Zhang, E. Hu, and A. Imamoglu: *A quantum dot single-photon turnstile device*. Science, 290(5500):2282 – 2285, December 2000.

[Mül83] Müller, Günter: *Supraleitende Niobresonatoren im Millimeterwellenbereich*. Dissertation, Bergische Universtität - Gesamthochschule Wuppertal, 1983.

[MN+05] Miller, R., T. E. Northup, , K. M. Birnbaum, A. Boca, A. D. Boozer, and H. J. Kimble: *Trapped atoms in cavity qed: coupling quantized light and matter*. J. Phys. B, 38:551 – 565, 2005.

[MRW88] Meystre, P., G. Rempe, and H. Walther: *Very-low-temperature behaviour of a micromaser*. Opt. Lett., 13(12):1078 – 1080, December 1988.

[MS92] Müller-Seidlitz, T.: *Untersuchungen am Ein-Atom-Maser*. Dissertation, Ludwig-Maximilians-Universität, München, 1992.

[MvdS99] Metcalf, Harold J. and Peter van der Straten: *Laser Cooling and Trapping*. Springer-Verlag GmbH, New York Berlin Heidelberg, 1999.

[MWM85] Meschede, D., H. Walther, and G. Müller: *One-atom maser*. Phys. Rev. Lett., 54(6):551–554, February 1985.

[NBFM93] Nez, F., F. Biraben, R. Felder, and Y. Millerioux: *Optical frequency determination of the hyperfine components of the $5s_{1/2} - 5d_{5/2}$ two-photon transitions in rubidium*. Optics Communications, 102:432–438, 1993.

[PBH+80] Pappas, P. G., M. M. Burns, D. D. Hinshelwood, M. S. Feld, and D. E. Murnick: *Saturation spectroscopy with laser optical pumping in atomic barium*. Phys. Rev. A, 21(6):1955–1968, June 1980.

[Pie73] Pierce, J. M.: *Methods of experimental physics*, volume 11. Academic Press, New York, 1973. p. 541.

[PP71] Panofsky, Wolfgang K. H. and Melba Phillips: *Classical Electricity and Magnetism*. Addison-Wesley Publishing Company, Reading, Massachusetts, 1971.

[PSS93] Padamsee, H., K. Shepard, and R. Sundelin: *Physics and accellerator applications of rf superconductivity*. Annu. Rev. Nucl. Part. Sci., 43:635–686, 1993.

[Pur46] Purcell, Edward Mills: *Spontaneous transition probabilities in radio-frequency spectroscopy*. Phys. Rev., 69:681, 1946.

[Rai95] Raithel, Georg: *Rydbergatome - vom klassischen Grenzfall zur quantenmechanischen Realität*. Habilitationsschrift, Ludwig-Maximilians-Universität München, 1995.

[Ram85] Ramsey, Norman F.: *Molecular beams*. Oxford Univesity Press, Walton Street, Oxford OX2 6DP, 1985.

[RBMJKW59] Rotenberg, Manuel, R. Bivins, N. Metropolis, and Jr. John K. Wooten: *The 3-j and 6-j symbols*. The Technology Press, Massachusetts Institute of Technology, Cambridge, Massachusetts, 1959.

[RBW95]	Raithel, G., O. Benson, and H. Walther: *Atomic interferometry with the micromaser*. Phys. Rev. Lett., 75(19):3446–3449, November 1995.
[Rem86]	Rempe, Gerhard: *Untersuchung der Wechselwirkung von Rydberg-Atomen mit Strahlung*. Dissertation, Ludwig-Maximilians-Universität München, 1986.
[RFH+03]	Rempe, G., T. Fischer, M. Hennrich, A. Kuhn, T. Legero, P. Maunz, P. W. H. Pinkse, and T. Puppe: *Single atoms and single photons in cavity quantum electrodynamics*. Coherence and Quantum Optics VIII, pages 241 – 248, 2003.
[Riz88]	Rizzi, Peter A.: *Microwave Engineering*. Prentice-Hall, Englewood Cliffs, New Jersey, 1988.
[RSKW90]	Rempe, G., F. Schmidt-Kaler, and H. Walther: *Observation of sub-poissonian photon statistics in a micromaser*. Phys. Rev. Lett., 64:2783 – 2786, 1990.
[RW87]	Rempe, Gerhard and Herbert Walther: *Observation of quantum collapse and revival in a one-atom maser*. Phys. Rev. Lett., 58:353 – 356, 1987.
[SGO08]	Sheng, D., A. Perez Galvan, and L. A. Orozco: *Lifetime measurements of the 5d states of rubidium*. Phys. Rev. A, 78(6), 2008. 062506.
[Sta05]	Stania, Gernot: *Messung von Ericson-Fluktuationen*. Dissertation, Ludwig-Maximilians-Universität München, 2005.
[TGH+09]	Thoumany, P., Th. Germann, T. Hänsch, G. Stania, L. Urbonas, and Th. Becker: *Spectroscopy of rubidium rydberg states with three diode lasers*. Journal of Modern Optics, 2009.
[The84]	Theodosiou, Constantine E.: *Lifetimes of alkali-metal-atom rydberg states*. Phys. Rev. A, 30(6):2881–2909, December 1984.
[THS+09]	Thoumany, P., T. Hänsch, G. Stania, L. Urbonas, and Th. Becker: *Optical spectroscopy of rubidium rydberg atoms with a 297 nm frequency doubled dye laser*. Optics Letters, 34(11):1621–1623, June 2009.
[TLP76]	Tuan, Duong Hong, Sylvain Liberman, and Jacques Pinard: *Detection and study of rb rydberg states*. Optics Communications, 18(4):533–535, September 1976.

BIBLIOGRAPHY

[TQGC99] Tan, Kok Kiong, Wang Qing-Guo, and Hang Chang Chieh: *Advances in PID Control*. Springer-Verlag GmbH, London, 1999.

[TS75] Townes, C.H. and A.L. Schawlow: *Microwave spectroscopy*. Dover Publications Inc., 180 Varick Street, New York, N.Y. 10014, 1975.

[VBWW00] Varcoe, B. T. H., S. Brattke, M. Weidinger, and H. Walther: *Preparing pure photon number states of the radiation field*. Nature, 403:743–746, February 2000.

[Wal04] Walther, H.: *Measurement of number states and phase diffusion using the micromaser*. Journal of Modern Optics, 51(6-7):933–943, April-May 2004.

[Wei78] Weissbluth, Mitchel: *Atoms and Molecules*. Academic Press, Inc, 111 Fifth Avenue, New York, New York 10003, 1978.

[Wil67] Wilks, John: *Properties of Liquid and Solid Helium*. Oxford University Press, London, 1967.

[WSB04] Wellens, Thomas, Vyacheslav Shatokhin, and Andreas Buchleitner: *Stochastic resonance*. Rep. Prog. Phys., 67:45–105, 2004.

[WVEB06] Walther, Herbert, Benjamin T. H. Varcoe, Berthold Georg Englert, and Thomas Becker: *Cavity quantum electrodynamics*. Rep. Prog. Phys., 69:1325 – 1382, 2006.

[WVHW99] Weidinger, M., B. T. H. Varcoe, R. Heerlein, and H. Walther: *Trapping states in the micromaser*. Phys. Rev. Lett., 82(19):3795 – 3798, 1999.

[YUM+03] Yoshikawa, Yutaka, Takeshi Umeki, Takuro Mukae, Yoshio Torii, and Takahiro Kuga: *Frequency stabilization of a laser diode with use of lifgt-induced birefringence in an atomic vapor*. Applied Optics, 42(33):6645–6649, November 2003.

[ZLKK79] Zimmerman, Myron L., Michael G. Littman, Michael M. Kash, and Daniel Kleppner: *Stark structure of the rydberg states of alkali-metal atoms*. Phys. Rev. A, 20(6):2251–2275, December 1979.

Die VDM Verlagsservicegesellschaft sucht für wissenschaftliche Verlage abgeschlossene und herausragende

Dissertationen, Habilitationen, Diplomarbeiten, Master Theses, Magisterarbeiten usw.

für die kostenlose Publikation als Fachbuch.

Sie verfügen über eine Arbeit, die hohen inhaltlichen und formalen Ansprüchen genügt, und haben Interesse an einer honorarvergüteten Publikation?

Dann senden Sie bitte erste Informationen über sich und Ihre Arbeit per Email an *info@vdm-vsg.de*.

Sie erhalten kurzfristig unser Feedback!

VDM Verlagsservicegesellschaft mbH
Dudweiler Landstr. 99
D - 66123 Saarbrücken

Telefon +49 681 3720 174
Fax +49 681 3720 1749

www.vdm-vsg.de

Die VDM Verlagsservicegesellschaft mbH vertritt

Printed by Books on Demand GmbH, Norderstedt / Germany